WORK AND HEALTH

Wiley Series On
ORGANIZATIONAL ASSESSMENT AND CHANGE

Series Editors:
Edward E. Lawler III and
Stanley E. Seashore

WORK AND HEALTH

Robert L. Kahn

Institute for Social Research
University of Michigan

A WILEY-INTERSCIENCE PUBLICATION

JOHN WILEY & SONS,
New York • Chichester • Brisbane • Toronto

ISBN 0-471-05749-5

Printed in the United States of America

10 9 8 7 6 5 4 3 2

For Sam and Jacob,
Jenny and Rebecca,
whose work awaits them . . .

Series Preface

The ORGANIZATIONAL ASSESSMENT AND CHANGE SERIES is concerned with informing and furthering contemporary debate on the effectiveness of work organizations and the quality of life they provide for their members. Of particular relevance is the adaptation of work organizations to changing social aspirations and economic constraints. There has been a phenomenal growth of interest in the quality of work life and productivity in recent years. Issues that not long ago were the quiet concern of a few academics and a few leaders in unions and management have become issues of broader public interest. They have intruded upon broadcast media prime time, lead newspaper and magazine columns, the houses of Congress, and the board rooms of both firms and unions.

A thorough discussion of what organizations should be like and how they can be improved must comprehend many issues. Some are concerned with basic moral and ethical questions— What is the responsibility of an organization to its employees?— What, after all, is a "good job"?—How should it be decided that some might benefit from and others pay for gains in the quality of work life?—Should there be a public policy on the matter? Yet others are concerned with the strategies and tactics of bringing about changes in organizational life, the advocates of alternative approaches being numerous, vocal, and controversial; and still others are concerned with the task of measurement and assessment on grounds that the choices to be made by leaders, the assessment of consequences, and the bargaining of equities must be informed by reliable, comprehensive, and relevant information of kinds not now readily available.

The WILEY SERIES ON ORGANIZATIONAL ASSESSMENT AND CHANGE is concerned with all aspects of the debate on how organizations should be managed, changed, and controlled. It includes books on organizational effectiveness, and the study of organizational changes that represent new approaches to organization design and process. The volumes in the series have in common a concern with work organizations, a focus on change and the dynamics of change, an assumption that diverse social and personal interests need to be taken into account in discussions of organizational effectiveness, and a view that concrete cases and quantitative data are essential ingredients in a lucid debate. As such, these books consider a broad but integrated set of issues and ideas. They are intended to be read by managers, union officials, researchers, consultants, policy makers, students, and others seriously concerned with organizational assessment and change.

The present volume, *Work and Health,* is distinctly different from those preceding in the series, and complementary to them. A pervasive theme in the series is the idea that organizations must be assessed and constructively changed with reference to a broad spectrum of values and purposes. In this book, one kind of criterion—health—is singled out for intensive treatment. While all aspects of health are considered, main attention is given to mental health, and to the role of work and work environments in the achievement by individuals of positive wellbeing. The author treats the meaning of work to people in different life circumstances, summarizes what is known about the characteristics of jobs that can affect health, and describes the sources and effects of occupational stress and strain. The final pages review alternative strategies for promoting healthful employment in ways compatible with organizational effectiveness. This overview is particularly timely because several lines of research appear now to converge toward the formulation of public policies and private programs to enhance mental health at work. The author has been a leading contributor to this research over a span of several decades.

EDWARD E. LAWLER, III
STANLEY E. SEASHORE

Ann Arbor, Michigan
July 1981

Preface

Research workers usually write for their colleagues, a fact that limits their audience but puts no limits on the use of esoteric language and notation. In this book I have attempted an exchange—less jargon for a wider readership.

The idea for such a book about work and health was first proposed to me by Elliott Liebow in 1976. He believed that enough was known about the complex relationship between work and health for someone to risk stating the major findings unpretentiously, and that I should do it. I am grateful to him for that, for subsequent encouragement, and for careful reading and criticism. I would like to acknowledge here the support of NIMH contract no. 278–76–34(SM).

Other colleagues whose comments were helpful to me are Ivar Berg, James House, Edward Lawler, and Stanley Seashore. The book is better for their careful review.

To my friends and co-workers in the Social Environment and Health Program of the Institute for Social Research I am especially grateful. John R. P. French, Jr., James House, Toni Antonucci, Charlene Depner, Robert Caplan, and I work together in this research domain, and I gladly acknowledge their influence. Marjorie MacKenzie, who is also a member of this research program, has brought to this book her careful bibliographic and secretarial work and her unfailing diligence and good humor.

Finally, I wish to thank the following authors and publishers for permission to quote or adapt material from their publications:

Studs Terkel and Pantheon Books (Random House), Alfred Slote and Bobbs-Merrill, and John Wiley & Sons.

Robert L. Kahn

Ann Arbor, Michigan
August 1981

Contents

WORK AND
HEALTH

CHAPTER ONE

Work in Industrial Societies

POINT OF DEPARTURE

Controversies about the meaning of work in human life are age-old and likely to persist. The human race has never been able to decide whether work is enslavement or liberation, curse or salvation, illness or therapy. I am not neutral in these matters and therefore state at the beginning the views that inform this book:

1.   Work is human activity that produces something of acknowledged value.

2.   All three elements in that definition are important for mental health—the fact and nature of the activity, the reality and experience of producing something, the recognition and acknowledgment by oneself and others that the activity and its outcome have value.

3.   In highly organized industrial societies work is available to most people only as jobs—ready-made packages of paid employment, so to speak. People generally do not create work roles that express their individual needs and abilities; each accepts the job available.

4.   Many jobs are deficient in one or another of the defining dimensions of work. The required activity may be trivial or meaningless, the thing produced may be remote from the worker, the value assigned by others may be meager. Such differences reduce or nullify the contributions of work to mental health.

5.   Whatever their adequacy on these defining dimensions, jobs have other characteristics important for illness or wellbeing. Among them are dependence or autonomy, powerlessness or control, isolation or belongingness, monotony or variety, danger or safety, overload and underutilization or goodness of fit.

6.   For most adults in industrial societies there is no satisfactory alternative to work; none within present aspiration and reach offers equivalent prospects for activity, meaning, recognition, and reward.

7.   It follows that the loss or denial of work is damaging to the individual and hazardous for the society.

These points lead to two major recommendations, each of them with many ramifications of policy:

Providing work is paramount for individual and social wellbeing; for this reason, jobs should be created as necessary.

The adequacy of jobs is almost as important as their availability; the quality of the work experience should therefore become a matter of national policy.

The early chapters of this book (1 through 7) offer some evidence for these conclusions; the latter chapters (8 through 10) examine the ways in which work and organizational life can be enriched.

THREE FACES OF WORK

Generalizations about work, and the statistical data on which they are based, are unavoidably removed from the events that occur when a unique individual encounters the requirements of a specific job. We rely on data for conclusions, but the words of individual workers illustrate more vividly the meaning of work in their lives. The following quotations are taken from interviews with three people who perform very different jobs. Work is important to all of them, although not for all the same reasons. The regularity, the sense of obligation, the enjoyment of contact with others at work, and the exchange of effort for economic reward are common to all three. Yet work presents to each a different face, and each is attached to work in ways somewhat different from the others. I call these ways affliction, addiction, and fulfillment, although each worker experiences something of each.[1]

Affliction

Here is Grace Clements, "a sparrow of a woman in her mid-forties," alienated from work but not from life. She is describing her job as a felter in a plant that manufactures luggage, where she has worked for 21 years. She makes the molded inner linings of suitcases.

> We're about twelve women that work in our area, one for each tank. We're about one-third Puerto Rican and Mexican, maybe a quarter black, and the rest of us are white. We have women of all ages, from eighteen to sixty-six, married, single, with families, without families.

> We have to punch in before seven. We're at our tank approximately one to two minutes before seven to take over from the girl who's leaving. The tank runs twenty-four hours a day.

> The tank I work at is six-foot deep, eight-foot square. In it is a pulp, made of ground wood, ground glass, fiberglass, a mixture of chemicals, and water. It comes up through a copper screen felter as a form, shaped like the luggage you buy in the store.

> In forty seconds, you have to take the wet felt out of the felter, put

[1]The following three interviews are excerpts from WORKING: People Talk About What They Do All Day And How They Feel About What They Do, by Studs Terkel. Copyright © 1972, 1974 by Studs Terkel. Reprinted by permission of Pantheon Books, a Division of Random House, Inc.

the blanket on—a rubber sheeting—to draw out the excess moisture, wait two, three seconds, take the blanket off, pick the wet felt up, balance it on your shoulder—there is no way of holding it without tearing it all to pieces, it is wet and will collapse—reach over, get the hose, spray the inside of this copper screen to keep it from plugging, turn around, walk to the hot dry die behind you, take the hot piece off with your opposite hand, set it on the floor—this wet thing is still balanced on your shoulder—put the wet piece on the dry die, push this button that lets the dry press down, inspect the piece we just took off, the hot piece, stack it and count it—when you get a stack of ten, you push it over and start another stack of ten—then go back and put our blanket on the wet piece coming up from the tank . . . and start all over. Forty seconds. We also have to weigh every third piece in that time. It has to be within so many grams. We're constantly standing and moving. If you talk during working, you get a reprimand, because it is easy to make a reject if you're talking. . . .

All day long is the same thing over and over. That's about ten steps every forty seconds about 800 times a day. . . .

I daydream while I'm working. Your mind gets so it automatically picks out the flaws . . . .You get to be automatic in what you're doing and your mind is doing something else . . . .

I hope I don't work many more years. I'm tired. I'd like to stay home and keep house. We're in hopes my husband would get himself a small hamburger place and a place near the lake where I can have a little garden of my own and raise my flowers that I love to raise.

Grace Clements fits well the definition of an alienated worker, engaging in activities that are not rewarding in themselves, that are demanding in some respects but permit little or no originality and discretion. Her work life is fragmented and machine-dominated, yet it is not without meaning. Her own work creates no complete product, but she contributes to the production of useful articles in ways that are visible to her. The economic value of her contribution is confirmed by the people who buy the luggage and by the wages she receives. She cares about the people she works with and represents them as head of the plant grievance committee: "They come to you angry, they come to you hurt, they come to you puzzled. You have to make life easier for them."

Addiction

When we speak of addiction to work, we use the term in a metaphorical rather than a literal sense, implying that some people are tied to their work as though to a drug—working more than required by reasonable social standards, perhaps working to a degree that compromises their performance as husbands or wives, parents and citizens. Such persons are often described as "workaholics," and we are occasionally told that as a nation we are "hooked on work."

I do not believe that the evidence justifies such assertions, although some jobs demand a tremendous investment of time and energy. Others may not demand such absorption, but they reward it and thereby motivate it. And some people are especially vulnerable to such demands and rewards. Ward Quaal, president of a large radio broadcasting corporation, describes his own well-rewarded and work-dominated life:

My day starts between four-thirty and five in the morning, at home in Winnetka. I dictate in my library until about seven-thirty. Then I have breakfast. The driver gets there about eight o'clock and oftentimes I continue dictating in the car on the way to the office. I go to the Broadcast Center in the morning and then to Tribune Square around noon. Of course, I do a lot of reading in the car.

I talk into a dictaphone. I will probably have as many as 150 letters dictated by seven-thirty in the morning. I have five full-time secretaries, who do nothing but work for Ward Quaal. I have seven swing girls, who work for me part-time. This does not include my secretaries in New York, Los Angeles, Washington, and San Francisco. They get dicta-belts from me every day. They also take telephone messages. My personal secretary doesn't do any of that.

She handles appointments and my trips. She tries to work out my schedule to fit these other secretaries.

I get home around six-thirty, seven at night. After dinner with the family I spend a minimum of two and a half hours each night going over the mail and dictating. I should have a secretary at home just to handle the mail that comes there. I'm not talking about bills and personal notes. I'm talking about business mail only. Although I don't go to the office on Saturday or Sunday, I do have mail brought out to my home for the week-end. I dictate on Saturday and Sunday. When I do this on holidays, like Christmas, New Year's, and Thanksgiving, I have to sneak a little bit, so the family doesn't know what I'm doing. . . .

I've always thought throughout my lifetime that if you have any ability at all, go for first place. That's all I'm interested in. . . . Sure I've been second vice-president, first vice-president, and executive vice-president. But I had only one goal in life and that was to be president.

As the head of a large corporation, Ward Quaal encounters work very different from that of Grace Clements, the assembly line felter. His work activities are varied, he travels extensively, he is involved in decisions that affect the amusement and education of a large public; he holds a coveted position and receives a large salary. Mr. Quaal may seem as driven by his dictating machine as Mrs. Clements is by her assembly line, but the comparison is deceptive. He has the power to delegate or to share the excessive burdens of his office, but he chooses not to do so. Work dominates his life, but we sense that he would be unhappy and troubled without it. As he says, "I am in a seven-day-a-week job and I love it!"

Fulfillment

Kay Stepkin is the director of a small nonprofit corporation, a bakery that produces and sells about 250 loaves of bread a day. She had been founder and owner of the bakery before it took its present form of organization. The variety of her activities, their direct and visible relationship to the product, and the values she shares with employees and customers make her work fulfilling.

I'm the director. It has no owner. Originally I owned it. We're a non-profit corporation 'cause we give our leftover bread away, give it to anyone who would be hungry. Poor people buy, too, 'cause we accept food stamps. We sell bread at half-price to people over sixty-five. We never turn anybody away. . . .

Everything we do is completely open. We do the baking right out here. People in the neighborhood, waiting for the bus in the morning, come in and watch us make bread. We don't like to waste anything. That's real important. We use such good ingredients, we hate to see it go into a garbage can. . . .

We have men and women, we all do the same kind of work. Everyone does everything. It's not as chaotic as it sounds. Right now there's eight of us. Different people take responsibility for different jobs. . . .

There isn't any machinery here. We do everything by hand. We get to know how each other is, rapping with each other. It's more valuable to hear your neighbor, what he has to say, than the noise of the machine. . . .

I get here at six-thirty, I stand at the table and make bread. I'll do that for maybe two hours. There might be a new person and I'll show him. . . . At eight-thirty or so I'll make breakfast and read the paper for half an hour. Maybe take a few phone calls. Then go back and weigh out loaves and shape 'em. . . .

Our prices are real reasonable. I went into a grocery store and saw what they were selling bread for. Machine-made whole wheat was selling at forty-five cents. So we made it fifty cents a loaf. It would cost fifty cents to make that bread at home using the same ingredients. . . .

I don't see how we'll ever make a profit because of the nature of what we're doing. There's a limit to how many loaves of bread one person could make. . . . But we're growing in other ways. We're looking for ways to get our product to people cheaper without resorting to machinery. One way is to get the ingredients cheaper—without sacrificing quality. So we've purchased a mill. It's going to make big changes here. . . .

We try to have a compromise between doing things efficiently and doing things in a human way. Our bread has to taste the same way every day, but you don't have to be machines. On a good day it's beautiful to be here. We have a good time and work hard and we're laughing. . . . I think a person can work as hard as he's capable, not only for others but for his own satisfaction. . . .

I am doing exactly what I want to do.

Work is an essential part of being alive. Your work is your identity. It tells you who you are . . . there's such a joy in doing work well.

(When people ask you what you do, what do you say?)

I make bread. (Laughs)

Kay Stepkin earns less money than Grace Clements and far less than Ward Quaal. Her economic rewards are slight, and she must cope with the risks that attend many small businesses in their early years. Her activities are varied, however, and her time is in some ways her own. She spends many hours at her work place but she has some leisure— "at eight-thirty or so I'll make breakfast and read the paper for half an hour." Her opportunities for conversation and interaction with others at work are almost unlimited. Moreover, her job comprehends to an unusual degree the many cycles of activity that link production and consumption. She and her co-workers choose their product, plan their work, make the complete product, and deal directly with the people who consume it. All these things are gratifying, and the gratification of work comes through in the interview.

WORK AS EXCHANGE

The societal exchange involved in working was epitomized long ago by Bertrand Russell (1930, p. 90): "In taking to agriculture mankind decided that it would submit to monotony and tedium in order to diminish the risk of starvation."

The transition from agricultural to industrial dominance, and now to the dominance of occupations that are neither agricultural nor industrial, has made the relationship of work to survival less direct, but the home truth remains. Moreover, individual employment also involves exchange. Holding a job is a contractual relationship, even though the contract is unwritten. The worker agrees to perform certain tasks, at certain times, in certain places, usually under the direction of certain persons. In exchange the employer or someone representing the employing organization makes a commitment of money.

The requirements of time and energy may be great or small, and they may reduce or enhance the worker's capacity for other activities. Work may be more dangerous or considerably safer than the person's nonwork pastimes. Work necessarily involves what economists call opportunity costs; the person who goes to a job foregoes the possibility of doing something else. How great those opportunity costs are depends on the pleasures foregone or postponed for the sake of work; they too may be great or small. If a man or woman considers work a rescue from boredom and unsuccessful attempts at amusement, the opportunity costs become negative and add to the attractiveness of the job. The rewards of the job itself, tangible and intangible, complicate the exchange still further, so that no single coinage can express it adequately.

Nevertheless, people understand the complex nature of work and have a good deal of insight into its characteristics, its demands and rewards, and the alternatives available to them. Some time ago, Robert Weiss and I asked several hundred employed men the following question:

> In your opinion, what makes the difference between something you would call work and something you would not call work?

Their answers emphasized the requiredness of work, the fact that it is paid, that it demands effort, and that it is productive. The first of these answers was the most common; once committed, you do your job whether you feel like it or not, and whether you like it or not. None of the answers implied that the exchange was necessarily bad; the majority of men said they would feel worse if they were not working. They used words like nervous, agitated, upset, guilty, ashamed, and bored to describe how they would feel without work (Weiss and Kahn, 1960).

These responses show that work gives more than money and that workers know it. Other studies lead to the same conclusion. At intervals over the past 25 years, we at the Institute for Social Research have asked nationwide samples of workers to answer the following question:

> If you were to get enough money to live as comfortably as you'd like for the rest of your life, would you continue to work?

Almost three-quarters of employed men and the majority of employed women say they would prefer to work even if they had no financial need to do so. That preference seems stable. It has not changed significantly during the past decade, and it is consistent with other things people say about their jobs. Seventy percent of all workers surveyed say they met some of their best friends at places they have worked. Even the minority of workers who say that they would quit if they could afford to tend to mention their co-workers when asked what they would miss most if they were not working. The majority who would continue to work explain that having a job keeps them from being bored (50 percent) and gives direction to their lives (16 percent).

Although work offers rewards other than money, it often demands more than task performance. The U.S. Department of Labor recognizes 19 areas of labor standards, and in five of them about 40 percent of all workers report problems: health and safety hazards, transportation, unpleasant physical conditions, inconvenient or excessive hours of work, and inadequate fringe benefits. On balance, however, the exchange is positive and workers know it.

WORK AND THE PERSON

Thinking of work as exchange clarifies a great deal about the relationship of worker to work, but the relationship goes deeper than exchange. It affects both the worker and the job. To some extent each person, having chosen a job, shapes its content. And to some extent the content of the job shapes the person. Not only are we affected by what we do; we tend to become what we do. A person's activities determine what has been called self-identity, and in our culture paid employment is the activity against which others are assessed.

When people ask that most self-identifying of questions— Who am I?—they answer in terms of their occupation: tool maker, press operator, typist, doctor, construction worker, teacher. Even people who are not working identify themselves by their former work or their present wish for it, describing themselves as retired or unemployed. And work that is not paid lacks significance, much as we might wish it otherwise. Many people who are usefully occupied, but not paid, respond to questions in ways that deprecate both their activities and themselves. A woman who takes care of a home and several small children and is engaged in a wide range of community activities may answer with that tired and inaccurate phrase, "just a housewife." A retired man, equally busy with an assortment of projects, is likely to say, "Oh, I'm retired; I don't do anything."

The majority of workers say that their job helps them understand the kind of person they really are and that a great many things in their lives depend on how well they perform on the job. When they are asked which of those other things are most important, they mention respect from others, happiness at home, and future career opportunities—even more often than being able to afford things they want to buy. It is consistent with these feelings about the importance of work in their own lives that people judge others by the work they do. The majority of men and women say that they can tell "something or a lot" (rather than "little or nothing") just from knowing what a person does for a living, and traits of personality or character are cited most often as examples.

Many theories deal with the basic question of how the situations an individual encounters and the activities that develop in those situations become part of person and personality. Theories of education, child development, and psychotherapy attempt to describe this process under different and special circumstances; more general theories of behavior try to deal with all situations and their effects in common terms. B. F. Skinner's (1967, 1974) theory of operant conditioning, for example, is built around the proposition that people will want and tend to behave in ways that have been rewarded in the past.

For our purposes, it is useful to think of a person's life as a set of roles, each with its own requirements and expectations, rewards and penalties. Thus a woman may have a job (worker), be married (wife), have children (mother), and so on. The list of major life roles may be long for some people, but it is not endless. In each role the person encounters and relates to other people, meets their expectations, more or less, and to some degree fulfills his or her own needs. In the course of such role behavior, repeated over time, the person is formed or altered; we can think of an identity, a part of the self, that corresponds to each of the major life roles. The employed man or woman thus develops an occupational self-identity, a sense of self as a worker and as a worker of a particular kind.

We can think of the person as having a core identity, as well as a set of interrelated subidentities that correspond to different roles. The requirements of a role come to be reflected in the corresponding subidentity, and the attributes of subidentities tend to be incorporated into the core identity. Thus the job of accountant calls for meticulous attention to detail; the accountant comes to think of himself or herself as appropriately careful of computations and tax regulations; and no one is surprised to find that the tendencies to check and recheck and to make sure that things are neat and orderly become characteristic of the person in general.

To the extent that a person's job offers a wide range of highly valued activities, its influences on the occupational subidentity and on the core self are positive. The research scientist's job calls for setting insightful problems and solving them; the opportunities to use highly valued skills and acquire new ones are many, and the general perceptions of oneself as problem solver and benefactor are gratifying. The person whose job requires a mindless repetition of a narrow set of activities, however, must cope not only with immediate boredom and frustration but also with more enduring implications for the self. How to devote time and energy to doing a mindless job without being considered—or worse, considering oneself—in some sense mindless is a difficult problem.

To the extent that work is regarded merely as exchange, a job can be assessed according to the time and energy it requires and the goods and services it provides to the worker. To the extent that the characteristics of the job affect the person and personality, the criteria for evaluating jobs are enlarged. The quality of employment, as well as the opportunity for employment, becomes an issue.

QUALITY OF EMPLOYMENT

Quality of work is a phrase with two meanings, and we have become accustomed to concentrating on only one of them—the quality of the outcome or product. As consumers that is our primary concern, usually within the constraint of price. It is also the primary concern of most supervisors: Are the people for whose work I am responsible doing good work?

Quality of work has an additional meaning, however; it can refer to the quality of the work experience. Both meanings are important, but they are not the same, as having a good job is not the same as doing a good job. I use *quality of employment* to refer to the things that make a job good or bad to have.

Two kinds of differences among individuals make the quality

of employment difficult to assess: differences in needs and abilities and differences in standards of judgment. A job that is too strenuous or too demanding of vigilance or isolating for one person may be just right for someone else. To some extent, therefore, the quality of a job must be assessed in relation to the needs and abilities of the individual who holds it. Goodness is not absolute, and we must think in terms of goodness of fit between person and job.

Moreover, judging the quality of anything implies judges, and here enters the second complication in assessing quality of employment: who is to determine whether a job is good or bad? One answer is obvious: the jobholder, who experiences the demands and rewards of the job directly and continuously. No observer can enter fully into that experience, and none should deny the relevance of the worker's own evaluation of the job. In a sense, the satisfaction or dissatisfaction of the jobholder is the ultimate judgment. If the worker finds the job to be drudgery, the contrary opinion of an observer will not alter the experience.

A statement of satisfaction, however, is a complex response; it says something about the thing being judged and something about the person doing the judging. People with limited experience tend to expect little and to be satisfied with little—more than they have perhaps, but little nevertheless. Most people come to terms with the realities of their lives, and their responses of satisfaction and dissatisfaction reflect those realities. When a man or woman answers "satisfied" in response to a question about the job, that answer has built into it the experience and expectations of the individual.

Strauss (1974) describes an interview with a blue-collar worker whose immediate response to the quality-of-employment question was, "I got a pretty good job." "What makes it such a good job?" the interviewer asked. The answer reveals something of what lies beneath such brief responses:

> Don't get me wrong. I didn't say it is a *good* job. It's an O.K. job—about as good as a guy like me might expect. The foreman

leaves me alone and it pays well. But I would never call it a good job. It doesn't amount to much, but it's not bad. (p. 55)

These complications do not make it impossible to determine the quality of employment, but they make it necessary to distinguish between objective and subjective measures. Prolonged noise at certain decibels and frequencies will damage a worker's hearing even if he or she reports no dissatisfaction with the noise level; another worker may be annoyed and dissatisfied with a noise level that is not physiologically damaging. Both kinds of data—the satisfactions and dissatisfactions of people on the job and the research findings and judgments of scientists—are relevant to measuring the quality of employment.

Proposals to improve the quality of employment quickly encounter the question of costs. There is no single answer to the cost question; some improvements in the quality of employment are costly, some are not, and some reduce costs. Between 1900 and 1940, the average workweek was cut from 60 hours to less than 40. That almost certainly improved the quality of employment and increased the costs of production, and the vast majority of Americans consider it a good exchange.

Some plants are now converting long assembly lines to other production methods, mainly to improve the quality of employment, but these conversions seem to involve neither consistent losses nor gains in production. One can imagine improvements in the quality of employment that might increase commitment and creativity, reduce waste, and thus reduce costs. The Japanese Quality Circles appear to be such an improvement, and others will develop. The costs and benefits of improving the quality of employment have only begun to be examined.

THE FUTURE OF WORK

Three counterarguments recur whenever proposals to improve the quality of the work experience are made: Such improvements are too costly; workers like things as they are, and modern technology is rapidly doing away with work anyway. The first two arguments have already been considered briefly; we turn now to the third.

In the years immediately following World War II, the introduction of automated machines and the increasing use of the computer generated visions of a world without work. Continuous-flow production methods, self-correcting machines, computer-instructed lathes and looms, and other such developments were cited as evidence of the end of work. This workless vision bears little relation either to present reality or, so far as we can see, to the reality of the future.

It fails to consider how needy most of the world's population is and ignores the limits to fossil fuels that are now so apparent. It confuses changes in technology and forms of work with the abolition of work. We may have entered a postindustrial era in the United States, in the same sense that we moved years ago into a postagricultural era. There is no implication that either agriculture or industry has become obsolete; rather, technological improvements have reduced the proportion of the population required for agricultural and industrial production. The proportion engaged in the creation and delivery of services—education, entertainment, medicine, and many others—has correspondingly increased.

In short, there is no danger that work will become unnecessary; the problem is to make it available and fulfilling.

SUMMARY

This book is committed to the importance of work and the improvement of the work experience. The availability and the quality of employment are proposed as national priorities. Most people want to work, not only for money but also for what work brings to their lives. They want to work in spite of the fact that many hold jobs that do not use their abilities well. The prediction that technology is making us a workless society is thus contrary to people's needs and wishes. Fortunately, it is also contrary to the emerging facts; reductions in agricultural and some branches of industrial employment have been accompanied by expansion in newer industries and in the services. The wise allocation of employment and the improvement of its quality emerge as key problems.

CHAPTER TWO

Occupational Differences in the Meaning of Work

What kind of work do you do? For whom are you working? What kind of business or industry is that? Census enumerators ask questions like these for every employed person in the United States. The answers are classified according to an index that was developed for purposes of the census and is revised as new jobs are created or old ones vanish.

It is a substantial work. The current index (The Alphabetical Index of Industries and Occupations, U.S. Bureau of the Census, 1971) contains 23,000 different job titles and 19,000 different kinds of business and industry. They are listed alphabetically, for the convenience of the reader, who can locate any needed entry or merely browse, from abalone processors and abnormal psychologists to zinc skimmers and zipper setters. The browser is certain to be impressed with the number and diversity of occupations and perhaps even more with the degree of specialization. The ancient art of making sausages, for example, is divided among sausage canners, sausage cookers, sausage cutters, sausage grinders, sausage linkers, sausage smokers, sausage stringers, sausage stuffers, sausage tiers, and sausage wrappers.

Census taking is a venerable activity. Even in so young a country as the United States, there has been a decennial census since 1790. Looking backward through the censuses, we find that the list of occupations has been growing. This growth is partly the result of invention, the creation of goods and services that did not previously exist. But it is partly the result of increasing specialization or fractionation of jobs. A statistician might say that the partitioning of the variance of work has been changing, so that more of the variance lies *between* jobs and less *within* them.

To put it another way, work comes in packages, each of them a collection of tasks or activities designed to be performed by one person. When we talk of getting a job, we mean getting entitlement to one of those packages. And from that entitlement much else in our lives is determined—what we do and where; how we are paid and how much; what vacation from work we may have and how often we may expect to be unemployed; where we are able to live; what we may be able to do for our children; how others see us and perhaps how we see ourselves.

OCCUPATIONAL PREFERENCES

People understand these things very well, and, in spite of the complexity of the occupational structure, they agree on which jobs are to be preferred. Social scientists may argue about *why* some jobs exceed others in material and nonmaterial rewards, but people clearly recognize *which* jobs are so advantaged. Since Counts (1925) first asked school children to arrange a list of 45 occupations in descending order from the one "most looked up to" to the one least admired, there has been a long procession of prestige-ordering occupational studies. In combination, they reflect a consensus that holds regardless of time, place, and characteristics of the rater.

Counts's work with school children was replicated at least four times over a period of more than 20 years, with findings of remarkable stability. Even years of depression and war made little difference in the ranking of occupations (Niez, 1935; Form, 1968). Prestige ratings of occupations by college students and adults of varying backgrounds agree with the earlier ratings by school children and show similar stability.

Cross-national comparisons are more difficult, because studies made in different countries used somewhat different lists of occupations. Inkeles and Rossi (1956), after adjusting for such discontinuities, reported a high consensus among six nations on the ranking of occupations common to all. The nations were the United States, Japan, Great Britain, the Soviet Union, New Zealand, and Germany.

All these studies depended on small groups of judges or raters. The variety of the raters and the convergence of the results suggest that the desirability ranking of occupations is widely shared, but only a representative sample of the population can allow such a conclusion to be made with confidence. Exactly such a study was done for the first time by the National Opinion Research Center in 1946 (Reiss et al., 1961). A sample of almost 3,000 adults throughout the United States was asked to evaluate 90 occupations according to their prestige. Not only was there general agreement on the prestige ranking; there was agreement also between the prestige ranking of occupations and the socio-economic gradations among them.

The best ratings were given to occupations that are well paid, "clean," require extensive and specialized training, and involve considerable responsibility for the welfare of others. Furthermore, they are jobs that include a number of tasks, challenging in their variety and complexity. Supreme Court Justice and physician led the list. At the other end were jobs that were unskilled, "dirty," and usually low paid—shoe shiner, street sweeper, garbage collector. It is perhaps not accidental that in recent years the occupation of shoe shiner has become rare, the sweeping of streets has become machine-assisted, and the collecting of garbage has become better paid and, by virtue of plastic bags and household garbage-grinders, somewhat less unpleasant. During the same period distrust in government increased greatly.

Whether these changes have altered the long-stable ranking of occupations remains to be learned. It seems likely that the rating of jobs as good or bad, which has held in so many times and places, will continue to be stable.

That stability comes partly from the fact that the preferential ordering of jobs is overdetermined; that is, it depends not on any one underlying characteristic, but on many. The jobs that are at the top of the list are not only better paid and more interesting in content; they provide more and better perquisites, steadier as well as higher income, more autonomy in choosing tasks and the methods for accomplishing them, and more control over one's own time. One can easily imagine small changes in specific rank occurring for historical or technological reasons, but the main ordering is deeply rooted. Table 2.1, in which the 10 top-rated jobs in the original list of 90 occupations compiled by North and Hatt (1962) are shown opposite the 10 rated at the bottom, illustrates vividly this combination of differentiating characteristics.

Table 2.1 Prestige Ratings for Selected Occupations[a]

Top-Rated Jobs	Bottom-Rated Jobs
U.S. Supreme Court Justice	Dock worker
Physician	Night watchman
State Governor	Laundry worker (presser)
Cabinet member	Soda fountain clerk
Diplomat, U.S. foreign service	Bartender
Mayor of a large city	Janitor
College professor	Sharecropper (farmhand)
Scientist	Garbage collector
Member of Congress	Street sweeper
Banker	Shoe shiner

[a]Source: P. K. Hatt and C. C. North, Prestige ratings of occupations, in S. Nosow and W. H. Form (Eds.), *Man, Work, and Society* (New York: Basic Books, 1962).

Hatt and North used a scale that ranged from a theoretical maximum of 100 points for a job that everyone rated excellent, to a minimum of 20 points for a job that everyone rated poor.

The actual range, since 3000 raters were of course less than unanimous, was from 93 to 33. The same rating procedure was repeated in 1963 with a smaller sample, and the ratings agreed very closely with those of the original study ($r = .99$). If the 90 occupations in the original list are grouped into a few familiar census categories, we see the occupational preferences of Americans in their most stable form (see Table 2.2).

Table 2.2   Prestige Ordering of Occupational Groups[a]

Classification	Average Score
Government officials	90.8
Professional and semiprofessional workers	80.6
Proprietors, managers, and officials (nonfarm)	74.9
Clerical, sales, and kindred workers	68.2
Craftsmen, foremen, and kindred workers	68.0
Farmers and farm managers	61.3
Protective service workers	58.0
Operatives and kindred workers	52.8
Farm laborers	50.0
Service workers (except domestic and protective)	46.7
Laborers (nonfarm)	45.8

[a]Source: P. K. Hatt and C. C. North, Prestige Ratings of occupations, in S. Nosow and W. H. Form (Eds.), *Man, Work, and Society* (New York: Basic Books, 1962).

EXPERIENCE AND SATISFACTION

Folklore is rich with sayings about the shallow attractions of the unfamiliar: Familiarity breeds contempt, distance lends enchantment, and the grass is always greener on the other side. Critics of the preferential ordering of occupations by large population samples have objected that most people have little direct knowledge about most jobs they are asked to rate. An alternative approach to understanding the significance of occupational differences is to ask people only about the jobs they know best—their own.

This has been variously done—most often by asking direct questions about satisfactions and dissatisfactions on the job, sometimes by asking people hypothetical questions about whether they would go on working if they inherited enough money to make work economically unnecessary, occasionally in still other ways (Hodge, Siegel, and Rossi, 1965). People have been asked what advice they would give young men or women choosing a line of work, what jobs they would want for their own sons or daughters, and what they would choose for themselves if they had the opportunity to begin again. Herzberg and his colleagues (1957) took the first step toward their own theory of work when they asked some 200 engineers and accountants to describe a time when they felt especially satisfied and a time when they felt especially dissatisfied with their jobs. The results of all these studies are not identical but they are generally consistent with the preferential ordering of occupations already described. The experience of people who do particular jobs confirms the impressions of people who observe those jobs from a distance.

The Second Time Around

When people are asked what kind of work they would choose if they could start over again, the proportion who say they would choose the kind of work they are now doing ranges from 16 percent for unskilled auto workers, to 93 percent for urban university professors. Answers to this question, as given in several different studies, are summarized in Table 2.3.

The range of response is perhaps the most important message of this table, but the discontinuities in the distribution are also significant. The sharpest break comes between professional and nonprofessional jobs—that is, between the professions and all other occupations, white-collar and blue-collar. There is an additional difference between white-collar and "working-class" (blue-collar) occupations in favor of the former, but it is not a complete break; the two distributions overlap somewhat. To put it another way, the ranking of jobs *within* the white-collar and

Table 2.3 **Proportions in Occupational Groups Who Would Choose Similar Work Again**[a]

Professional and Lower White-Collar Occupations	Percent-age	Working-Class Occupations	Percent-age
Urban university professors[b]	93	Skilled printers	52
Mathematicians	91	Paper workers	42
Physicists	89	Skilled auto workers	41
Biologists	89	Skilled steelworkers	41
Chemists	86	Textile workers	31
Firm lawyers[b]	85	Blue-collar workers	24
School superintendents[c]	85	Unskilled steelworkers	21
Lawyers	83	Unskilled auto workers	16
Journalists (Washington correspondents)	82		
Church university professors[b]	77		
Solo lawyers[b]	75		
White-collar workers[b]	43		

[a]Data in this table are based on responses to the question: "What type of work would you try to get into if you could start all over again?" Entries are primarily from a study conducted by the Roper organization, of 3000 workers in 16 industries.

[b]Probability samples or universes of six professional groups and a cross section of the "middle class" (lower middle class and upper working class) in the Detroit area, stratified for comparability with respect to age, income, occupational stratum, and other characteristics. From Wilensky (1964).

[c]From a 1952–1953 Massachusetts sample taken from Gross, Mason, and McEachern (1958).

working-class occupations shows differences at least as great as those *between* the white-collar and working-class groups.

The hypothetical question about "starting all over again" was asked only of people in selected industries, and therefore the full range of occupations is not represented. The response pattern, however, is familiar. People in the high-prestige occupations would choose them again: people in the low-ranked occupations, especially the unskilled blue-collar jobs, would choose differently if they had a chance to do so.

Work or Quit

"If you were to get enough money to live as comfortably as you'd like for the rest of your life, would you continue to work?" The answers to that question provide another measure of satisfaction with work and attachment to it. The question was first asked of a national sample of employed men and women in 1953: it was asked again in 1960, 1969, 1973, and 1977. During a period of 24 years, interviewers of the Survey Research Center (University of Michigan) have put that hypothetical question five times to employed men and women throughout the country. Their answers over the years show an interesting pattern of stability and change. To begin with, those who say they would go on working are a large majority—67 percent in 1969 and 1973, and 71 percent in 1977. Comparisons with earlier years cannot be made with the same confidence, but the proportion who chose work in answer to our question was at least as large. The overall stability is great.

Men and women differ in their answers, with more employed women than men saying that they would quit their jobs if they had enough money to do so. More than 40 percent of employed women give that answer, and the corresponding percentage for men has always been less than 30. This relative readiness of women to leave their jobs occurs almost entirely among those for whom work is a secondary or part-time activity. The work attachments of career men and women are very similar.

The *reasons* for these choices are also stable and in a sense saddening. Most people who would choose to continue working say that work keeps them from being bored or gives direction to their lives. People talk about making the time pass, of not knowing what to do with themselves if they were not working; a few say they would go crazy if they were not working. In contrast, among the workers who said they would choose to go on working, only one in 10 gave "enjoys working" as a reason, and the proportion who spoke of liking the work itself was even smaller.

Occupation enters into these decisions in two ways: It affects the decision to go on working at all, and it affects the kind of work people would choose to do. The proportion of people who say they would work even if they had no economic need to do so is highest among those in the jobs that stand high in the preferential ordering of occupations and is lowest among people who hold jobs near the bottom of that list. In 1973 the percentage who said they would choose to work was 75 among those in the professions and 78 among managers. The corresponding percentage for machine operators was 54.

Even more revealing are the answers to a question less often asked: If you did decide to go on working, what kind of work would you do? Here the answers show a pattern very like that evoked by the question about starting one's adult life over again. People in the professions, especially the most advantaged of them, would not merely "go on working," they would choose the job they already hold. People in jobs at the low end of the preference ordering, by contrast, seldom say they would go on working at their present jobs. Their dominant preference is for self-employment. A 48-year-old mail carrier says, "I'm thinkin' about goin' in business for myself." Grace Clements, the felting machine operator quoted in Chapter 1, ended her interview on the same note: "We're in hopes my husband would get himself a small hamburger place." An order filler in a shoe factory says, "I would like to work with children, small children. I have thought several times of trying to set up a nursery, even if I didn't start with but one or two children" (Terkel, 1972.)

The owner of a small factory answers the same question in a very different way: "Retire? Hell, no. I'd open up another shop and start all over again. What am I gonna do? Go crazy? I told you I love my work . . . . I don't have to worry about tomorrow. But I still want to work. I *need* to."

The evidence for these differences is quantitative as well as qualitative, as Morse and Weiss (1955) demonstrated years ago. In their national survey on the meaning of work, the question of whether people would work even if they inherited money was followed with another one: Would you still keep on doing the same type of work you are doing now?

The occupational contrasts in the responses to this pair of questions are illuminating. The majority of employed men, regardless of present occupation, said they would continue to work; the proportion giving that answer ranged from 58 percent of unskilled workers to 86 and 91 percent of professionals and salesmen. But the proportion who said they would continue in the same type of work varied much more—from only 16 percent of the unskilled workers to 68 percent of the professionals. The attachment to work is stronger and more general than the attachment to specific jobs. Only as we turn to jobs well up in the preferential ranking of occupations do we find a majority of people saying that they would choose to continue at the kind of work they are now doing.

These occupational differences help explain an apparent inconsistency between what people say and what they do. Most people say they would work even if they didn't need the money, but some early retirement options have been taken up quickly by the majority of eligible workers. The "30-and-out" agreement negotiated by the United Auto Workers and the major automobile manufacturers is a case in point. The large proportion of workers who chose "out," even at reduced pension amounts, surprised both the employers and the union.

The answer lies with occupational differences and choice. People whose jobs are low in the preferential listing of occupations want to continue working, but not at the jobs they have. When the choice is between the present job or out, they choose out. How many of them subsequently realize their hope for work of other kinds is simply not known.

The Satisfaction Riddle

Direct questions about job satisfaction evoke the familiar pattern of occupational differences, but the pattern is muted because most people call themselves satisfied. If they are given no middle ground, almost 9 out of 10 people say they are satisfied rather than dissatisfied (Quinn and Shepard, 1974). Moreover, the tendency toward reported job satisfaction has been consistent over many years, in spite of substantial changes in the economy and the labor force. The reason for the pervasiveness and stability of this attitude is that a "satisfied" response is a complex reaction. It says something about the situation and something about the person who is responding to it. It involves many comparisons, many frames of reference, many hypothetical alternatives. This is what the blue-collar worker was trying to explain to the interviewer in the conversation quoted in Chapter 1. The interviewer's first question, about what appeared to be very routine work, produced a one-sentence answer that would have been coded "satisfied": "I got a pretty good job."

It was the second question that clarified the situation:

What makes it such a good job?

Don't get me wrong. I didn't say it is a *good* job. It's an OK job—about as good as a guy like me might expect. The foreman leaves me alone and it pays well. But I would never call it a *good* job. It doesn't amount to much, but it's not bad.

Occupational differences in overall job satisfaction reflect very real differences in conditions of work, modified by the considerable ability of human beings to come to terms with such differences. In the nationwide studies of the Survey Research Center, "facet-free" job satisfaction was measured using five general questions about workers' attitudes toward their jobs. These were scored on a scale from 1.00 to 5.00, with higher scores indicating greater satisfaction (Quinn and Shepard, 1974).

Table 2.4 shows the satisfaction scores for different occupational groups, along with their ranking.

Table 2.4   Job Satisfaction in Relation to Occupation

Occupation	Score	Rank
Professional and technical workers	4.11	1
Managers and administrators	4.00	2
Crafts workers	3.89	3
Salespersons	3.82	4
Clerical workers	3.67	5
Machine operators (except transport)	3.39	6
Laborers (nonfarm)	3.28	7

Three things stand out in this table. First, the range is small; on a scale that can reach from 1 to 5, the mean scores for the different occupational groups range only from 3.28 to 4.11. Second, the entire distribution is skewed toward the "satisfied" end of the distribution; most people report themselves satisfied to some degree, rather than dissatisfied. Third, the rank order of satisfaction scores corresponds very closely to what we have seen in the prestige ordering of occupations. Even the slight margin of crafts workers over sales and clerical workers is similar to the prestige overlap between certain white-collar and blue-collar occupations seen in Table 2.3.

OCCUPATIONAL EFFECTS ON HEALTH

Folk wisdom identified some occupational hazards long before toxicology, industrial medicine, or occupational mental health had been heard of. The phrase "mad as a hatter" came into the English language long before we knew the effects of mercury, which was used in felting, on the central nervous system. And so it has been with other occupations and diseases, among them the "black lung" of coal miners and the melancholia of sailors.

In more recent years the documentation of such occupational problems has become extensive (American Public Health Association, 1975), and general legislation for occupational health and safety has been enacted. Research on the causes of job-related physical disability is well advanced; research on the mental and emotional consequences of work is beginning to be seriously regarded (Institute of Medicine, National Academy of Sciences, 1981).

Physical Hazards

Much of that research has developed along occupational lines. Categories of physical illness associated with different occupations are usually discovered first, and research workers then attempt to identify the physical substances and kinds of exposure that account for such associations. This research strategy has already generated a great deal of information, and it continues to do so.

For example, workers who mine or process asbestos in the United States have about five times the rate of fatal lung cancer expected in the general population, and the ratio of actual to expected deaths increases with exposure (Selikoff and Hammond, 1975). Twelve percent of all working coal miners and 21 percent of nonworking miners showed X-ray findings of pneumoconiosis (Leinhart, Doyle, et al., 1969). Women workers exposed to anesthetic gases in operating rooms give birth to infants with congenital abnormalities about twice as frequently as similar workers not so exposed (American Society of Anesthesiologists, 1974). Meat wrappers using polyvinyl chloride materials have distinctive respiratory symptoms (69 percent) and characteristic remissions during weekends and vacations (77 percent) (Bardand, 1973).

Since the passage of the Occupational Safety and Health Act in 1970, standards for the use of these and other materials have been rapidly established. At least 25 such standards were set in the period of 1972 to 1975 for stressors ranging from the commonplace to the exotic—from heat and noise to beryllium and toluene diisocyanate. Much remains to be done, but the process of discovery and application is well advanced with respect to the physical hazards of work. It is still new with respect to social and psychological stresses, and it has just begun with respect to the positive aspects of work.

Psychosocial Indicators

As with the studies of work and physical health, the early research linking work to problems of mental health is based on differences among occupations, industries, and classes. The incidence of mental illness was found to increase in successively less advantaged social classes, and the patterns of diagnoses differed as well for people in different classes. Schizophrenia, for example, was diagnosed more frequently among people in the lower socioeconomic classes (Hollinghead and Redlich, 1958).

A study of mental health and illness in New York City (Srole, Langner, et al., 1962) used a 6-point scale ranging from well (no symptoms) to incapacitated. The resultant "sick–well ratios" increase steadily in successively lower socioeconomic categories, from 46 in the top category to 470 in the bottom, more than a tenfold difference. Moreover, these effects are aggravated by downward mobility. The sons of professional and executive fathers who did not themselves enter such occupations show 20 times the symptoms (sick–well ratio) of sons who followed their fathers into these advantaged occupations.

Reported Problems

In more recent studies (Quinn et al., 1974; Quinn, Walsh, and Hahn, 1977) nationwide samples of men and women were interviewed about their work situations, with special emphasis on 19 aspects of working conditions that have been the subject of government action, legislation, and labor negotiation. Nine out of ten workers reported at least one problem in one of these 19 areas, and the average number of problems mentioned was three. Behind that average lies a familiar pattern. More problems were mentioned by wage-and-salaried workers than by the self-employed, more by blue-collar than by white-collar workers, and more by machine operators than by those in any other occupation.

This general finding of disadvantage in blue-collar, industrial, and service jobs summarizes an almost unbroken pattern of specific disadvantages. Reported exposure to physical danger, for example, is more than twice as common among blue-collar workers (crafts workers, machine operators, transport operators) as among professionals, administrators, and salespeople. Less obvious but no less significant are the occupational differences in employment security. Unsteady employment is essentially a problem of certain occupations and industries. It is reported by only 1 to 3 percent of white-collar workers in government, finance, insurance, and real estate; but it is reported by 40 percent of construction workers, 20 percent of miners, and more than 20 percent of unskilled workers regardless of industry.

There are some exceptions to this pattern. Complaints about transportation and traffic are more frequent among white-collar than blue-collar employees. The flight to the suburbs imposes costs on those who move as well as on those who do not. In general, however, the pattern of advantage and disadvantage is consistent rather than compensatory; disadvantaged jobs are consistently disadvantaged. Quinn and Mangione (1977) compared the major occupational groups with respect to seven kinds of employment benefits—medical insurance, paid vacations, life insurance, retirement pensions, paid sick leave, profit sharing, and stock options. In spite of the fact that some of these benefits have come to be routinely included in contract negotiations between labor unions and employing organizations, all these benefits are more characteristic of professional and managerial jobs than of machine-operating or service jobs. Paid sick leave, for example, goes with 95 percent of professional and technical jobs, but with only 34 percent of machine-operating jobs.

Attitudes and Behavior

The pattern of occupational differences that has by now become familiar does not extend to most measures of on-the-job behavior that were available—self-reported lateness to work, absence from work, stated intention to quit, and the like. It is likely that these behaviors are under the control of organizational rules and penalties for tardiness and absence and that decisions to quit or remain on a job depend more on the labor market and the available alternatives than on present satisfaction.

Only the making (or not making) of suggestions for changes to one's employer revives the pattern of occupational differences, and with a reversal that is both plausible and ironic. People in occupational groups who describe their jobs as having the most problems and themselves as being least satisfied nevertheless make the fewest suggestions to their employers. Laborers and machine operators are at the bottom of this frequency distribution; professionals and managers are at the top. Some of this difference no doubt reflects lack of expertise. Much of it probably reflects limited access to information, skepticism about the receptivity of management to suggestions from the rank and file, and lack of commitment to the job.

Occupations and Ramifications

The feelings that people have about their work extend to other domains of life. In the 1973 study of work (Quinn and Shepard, 1974) an index of life satisfaction was constructed from 10 questions, two dealing with the person's overall feelings of happiness and satisfaction and eight dealing with the person's more specific feelings about life—whether it was interesting or boring, worthwhile or useless, hopeful or discouraging, and the like. The index of life satisfaction, which gave equal weight to the person's general and specific feelings, varied from -24 to $+23$. Machine operators were at the bottom of that range, and professionals were at the top.

Campbell, Converse, and Rodgers (1976), in another nation-wide study, provide further evidence for the ramifying effects of occupational differences. They developed separate scores of sat-isfaction for 17 "domains of life experience," such as marriage, family, health, neighborhood, job, and housing. They then inves-tigated the extent to which people's general sense of well-being depended on their satisfaction in each of the 17 domains of expe-rience.

Table 2.5   Proportion of Variance (r^2) in Index of Well-Be-ing, Explained by Individual Domain Satisfaction Scores[a,b]

	Proportion of Explained Variance (r^2)
Nonworking activities	29
Family life	28
Standard of living	23
Work	18
Marriage	16
Savings and investments	15
Friendships	13
City or county of residence	11
Housing	11
Amount of education	9
Neighborhood	8
Life in United States	8
Usefulness of education	8
Health	8
Religion	5
National government	5
Organizations	4

[a]From A. Campbell, P.E. Converse, and W.L. Rodgers, *The Quality of Ameri-can Life* (New York: Russell Sage Foundation, 1976), p. 79.

[b]The number of cases on which each estimate of explained variance is based ranges from 2106 to 2160, with three exceptions: family life (2077 cases), amount of education (1975 cases), and organizations (1783 cases).

As they put it (p. 77): "Granted that satisfactions and dissatisfactions must combine in some way to influence global reports of well-being, how do our respondents go about this process of amalgamation and summarization?" The answer, after a number of more complicated possibilities were investigated, is that people's "assessment of their lives as a whole can best be explained as a weighted summation of their degree of satisfaction with the various specific domains." In effect, people assess the overall satisfactoriness of their lives by adding up the satisfactoriness of the different parts or domains, but not equally—they consider some parts of their lives more important than others.

If we list the 17 domains in order of their importance for overall satisfaction with life, a person's job and standard of living (which depends directly on the job) are high on the list. Only family life and all nonworking activities taken together are more important (Table 2.5).

SUMMARY

In short, work is important. People tend to agree about which jobs are good, regardless of what jobs they have themselves. However, their feelings about their own work, their hopes for the future, and their health, depend to a considerable extent on the kind of work they do. Satisfaction with work and satisfaction with life tend to go together.

CHAPTER THREE

The Content of Jobs

People who work often have some choice among jobs, but they don't have much choice about what goes into them. The job comes ready-made, with predetermined hours, rewards, places of work, people to work with, and demands for skill and strength. To some extent each man or woman puts a personal stamp on the job, but the characteristics of the job are there before the person arrives and they don't change much.

Jobs that have identical or similar characteristics can be thought of as a set, and such sets we call occupations. People think in these terms, as we have seen, and they have strong feelings about different occupations. People agree on the preferential ordering of occupations—which ones are good and which bad, to put it bluntly—and these preferences are remarkably stable; neither the passage of time nor the crossing of national boundaries seems to affect them much. Moreover, the feelings people have about their own jobs—their satisfactions and dissatisfactions, their wishes to continue or quit, their sense of what they would do if they could begin their working lives again—are consistent with their preferential ranking of occupations in general.

This pattern of research findings leads to an obvious but difficult question: Exactly what makes one job preferable to another? The question is important for both scientific and pragmatic reasons. From the scientific view, it is a specific example of a general question that is asked repeatedly: Why? Almost every research finding of significant interest, once it has been accepted, is subjected to that question. If some kinds of mental illness are more common among people of low socioeconomic status than among the prosperous, why? If coronary disease is more prevalent among physicians in general practice than among dermatologists, why?

Such questions are demands for additional information about some fact or relationship already discovered. They ask what it is that explains the relationship—what it is about low socioeconomic status that causes mental illness, for example, or what it is about general medicine that makes its practitioners more coronary-prone than dermatologists. In the process of answering such questions, sequences or patterns of scientific explanation are developed that satisfy us, logically and intuitively. The mark of such satisfaction is that, at least for a time, we ask no more about that particular relationship and turn instead to other inquiries.

THE PRAGMATIC QUESTION

Identifying those underlying characteristics that make jobs good is as important for pragmatic as for scientific purposes. Suppose that we want to make all jobs good. To make all of them equally good is perhaps too naive a goal to serve as an example, so let us suppose that we want simply to reduce the very wide range that now separates the most preferred from the least preferred occupations. We cannot achieve this goal by arranging for everyone to work in the top 10 occupations and abolishing the bottom 10. We cannot become a nation merely of Supreme Court justices, physicians, scientists, and professors.

We can improve jobs now ranked low by designing into them at least some of those underlying properties that make the preferred jobs so attractive. But this strategy brings us again to the scientific question of what those underlying properties really are. In this respect the scientific and pragmatic questions converge. The factors or variables that explain job satisfaction in scientific terms should also be, to a considerable extent, the things that must be changed in order to increase job satisfaction.

Of course, satisfaction is not the only desired outcome. Important though it is, satisfaction is not synonymous with health; and job factors that affect health may not affect satisfaction—at least until the worker's health is damaged. Moreover, such key job behaviors as productivity, absence, and turnover are likely to have some determinants quite separate from those that cause satisfaction. What job characteristics we conceptualize and measure depends on what our purpose is, what we are trying to predict and explain. Recruiters for the basketball team and members of the academic admissions committee are not likely to describe potential college students in the same way. The Sierra Club and a company of oil geologists use different concepts to describe the same mountain. *What* we single out for description about a thing depends on *why* we are trying to describe it, and people who have described jobs have done so for many different reasons.

Competing Approaches

There is no end to the ways jobs can be described and probably no one way that describes them completely and perfectly. Thousands of articles have been written on the subject of job satisfaction alone. Too few of them present significant amounts of empirical data, and all too many of them describe jobs in terms that are unique to the author and go unvalidated and unreplicated. Nevertheless, there are a few theoretical approaches that have been used repeatedly in empirical work. They include Likert's (1961) characterization of jobs and organizations on a set of related dimensions that range from authoritarian to participative; Maslow's (1954) theory of a need hierarchy, which leads each person to strive for the fulfillment of successively higher and more complex needs, from subsistence to self-actualization; Herzberg's (1959) proposal that some job characteristics create satisfaction and that others (not simply the absence of the "satisfiers") create dissatisfaction. In addition, over a period of 25 years there have been several serious reviews of the literature of job characteristics and job satisfaction (Brayfield and Crockett, 1955; Herzberg, Mausner, et al., 1957; Smith, Kendall, and Hulin, 1969; Porter and Steers, 1973; Srivastva et al., 1975; Locke, 1976).

These have been supplemented by a compendium of available measures of occupational attitudes and characteristics, compiled by Robinson and his colleagues (1969) at the Survey Research Center; and two standardized sets of job and organizational measures, the Survey of Organizations (Taylor and Bowers, (1972) and the Michigan Organizational Assessment Package (1975).

Quinn and his colleagues (1974, 1977) have developed measures of job characteristics and worker attitudes and have used them in successive national surveys.

We thus have available to us a number of theories about the underlying characteristics of jobs and several sets of measures, some of which derive from those theories. If the scientific appraisal of these various approaches were complete, we would also have careful comparisons of the agreements and disagreements among them and of the extent to which each set of job descriptors correctly predicts the outcomes of interest. These include the satisfaction or dissatisfaction of jobholders, their behavior in the work situation, their health, and the quality of their lives. The study of people at work illustrates this process of scientific comparison and evaluation, but only in its early stages.

EIGHT KEY CHARACTERISTICS

Occupations have been studied for a variety of purposes, from profit to self-actualization, and they have been studied by many kinds of investigators, from remote scholars to personnel analysts and time-study engineers. Each kind of investigator and investigation has distinct ways of going about the descriptive task, and the result is a profusion of questionnaire items and scales, observational categories, and other information-getting devices. Systematic tests for reliability and validity are rare, and the healthy pruning that such tests will impose has yet to be accomplished.

Despite these problems, a review of the accessible universe of job measures shows us a core of agreement. A few underlying properties appear recurrently in studies of work. The specific measures used to get at those properties vary a good deal, but the concepts appear repeatedly. The different purposes for which studies are undertaken are expressed more by the relative weight given to these different job components than by the choice of the components themselves.

Eight aspects or components of jobs constitute this set:

1. *Task content.*   Task content refers to what the person actually does on the job—whether the worker operates a machine that stamps out small metal parts, conducts interviews with applicants for clerical work, or makes deliveries from warehouse to supermarkets. This area is of interest to social scientists, is the main area of emphasis in the personnel specialty called job analysis, and is similarly emphasized, at a microscopic level, in time-study engineering. More detailed measures in this area include scales of variety or monotony, the length of the work cycle, and the simplicity or complexity of the task and the length of time required to learn it.

2. *Autonomy and control.*   At the task level, autonomy has been most often measured in terms of the worker's choice (or lack of it) regarding methods of work, the tools to be used, the sequence in which parts of the task are performed, and the pace at which work is accomplished. Control in this context has usually meant the control of the worker over his or her own time and activity, not control over others or participation in the general

control of the enterprise or its parts. The range is effectively set at one end by the machine-paced assembly line and at the other by the professional job in which the immediate goals and the means for attaining them are under the control of the individual jobholder.

3. *Supervision and resources.* Managerial and supervisory behavior may be the most nearly universal component of job description. Work organizations, even those that are democratic in a legislative sense, are hierarchical to some degree in their administration. No degradation of the democratic principle is necessarily implied. Organizations are networks of considerable interdependence, and the successful organizational outcome requires individual performance that meets certain standards and is coordinated with the activities of others.

It is a rare organization, therefore, in which each person's work is not somehow subject to review and supervision by others. The qualities of that supervisory process are almost always important, especially for the person supervised. It is for these reasons that supervision is considered a defining component of jobs, regardless of the theoretical preferences or pragmatic purposes of the investigator.

4. *Relations with co-workers.* Few people work alone; most are involved with others in their work, depend on others to get their own work done, and are depended on in similar fashion. The degree of interdependence varies, of course. In some cases a group of people must work closely together to turn out a single product. In other cases the members of a work group are less interdependent. Nevertheless, workers who are supervised by the same person usually are engaged in related tasks, work under similar conditions, and receive similar wages.

Such work groups can develop great power, informal as well as formal, over their members. The feelings of a person about his or her job are affected by the characteristics of the group and the person's relationship to it.

5. *Wages.* People do not work only for extrinsic rewards, but such rewards are extremely important. For most workers, the

payment for work is the sole significant source of income. The material aspects of life outside the job are thus determined by the material rewards of the job. In addition, the amount of money earned has great symbolic value; it is a means of communicating esteem. Finally, wages constitute a means of comparison; individuals compare themselves with others in terms of wages received, and questions of equity become scarcely less important than questions of adequacy. Most attempts to measure significant aspects of jobs include the amount of pay, and some include matters of comparison and equity as well.

6. *Promotion.*   The importance of promotions, like that of wages, is both practical and symbolic. In a hierarchical organization, promotion is often regarded as a recognition of past performance and an organizational vote of confidence for the future. For some people, especially at early stages of their careers, promotion and the likelihood of promotion may be more important than current wages. Nevertheless, promotions are measured less frequently than pay, perhaps because they are thought of as properties of a person's history or career rather than as characteristics of a job.

People understand that some jobs are a dead end and that others are stepping-stones to better things. It is possible to measure the promotive potential of jobs as well as the past promotions and prospects of individuals.

7. *Working conditions.*   The physical conditions of work can be measured objectively and subjectively. Space, freedom of movement, noise level, temperature, and ventilation are often measured, and statutory standards exist for some of these characteristics of the work environment. Subjective measures have also been used, most frequently in the form of questions about comfort.

8. *Organizational context.*   In addition to the seven aspects of jobs summarized above, some investigators have attempted to measure the organizational context or structure or climate in which the job exists. The size of the organization, for example, is

such a factor, as are the number of echelons or hierarchical levels, public or private ownership, and organizational product.

It can be argued that these organizational properties, although important in their effects, are not aspects of jobs. If they affect jobs, the argument continues, they must do so by altering one or more of the previous seven factors. This is plausible, but until such causal sequences have been established, it is perhaps better to measure too much than too little. One feels intuitively that two jobs may be much the same in their immediate demands and rewards and yet experienced differently because of differences in the larger organizational context.

The things that make work important to people and the effects, good and bad, that work has on the jobholder can be described according to these eight aspects of jobs. Each of them can be further differentiated, of course. We can measure relations with co-workers, for example, with a single scale ranging from excellent to intolerable; but we can also imagine many scales, each dealing with some aspect of co-worker relations. Such elaborations of measurement, however, fit comfortably within one or another of the eight major aspects of jobs. One of the more ambitious examples of such measurement is the Quality of Employment Survey. It illustrates also how detailed measurements can be combined to increase our understanding of what makes jobs good or bad in the eyes of workers.

THE QUALITY OF EMPLOYMENT SURVEY

Imagine that you had been chosen as one of the 2000 respondents for the Quality of Employment Survey. You have been called on by an interviewer, who is identified as a staff member of the Survey Research Center of the University of Michigan. You have listened to a well-rehearsed explanation of the importance of finding out not only whether people who want jobs have them, but also what the qualities of those jobs are. You have agreed to participate, and a few questions have been asked in the conventional way.

Now the interviewer takes out a deck of cards and says:

> The next question involves things a person may or may not look for in a job. Some of these things are on this set of cards. People differ a lot in terms of which of these things are more important to them. We'd like to know how important each of these things is to *you.*

The interviewer then puts down side by side four "alternative cards," each of which contains the beginning of a sentence. The first one reads: "It is *very important* to me to have a job where . . . ." The others read: "It is *somewhat important* to me to have a job where . . . ," "It is *a little important* to me to have a job where . . . ," and "It is *not at all important* to me to have a job where . . . ."

The interviewer then hands the entire deck of cards to you and says:

> Each of the cards in this deck contains a different statement about a job, something that some people might consider important and others might not. Please read each card and put it *below* the alternative card which best reflects *how important* each thing is to you.

You look at the first card in the deck, which reads, "I have enough time to get the job done." You think about how important having enough time at work really is for you and therefore below which of the four alternative cards you should put this one. The sorting process has begun.

When you finish sorting the cards, the interviewer carefully picks them up, keeping each stack of cards with the alternative card under which you had put them, and explains that they will be read by a computer and combined with the classifications made by other people throughout the country. Taken all together, the results will show what American workers think are important about jobs.[1]

Factor analysis was the statistical procedure chosen for analyzing workers' ratings of importance. It is a procedure that tells us whether certain ratings go together, whether there are clusters of ratings so closely related that they seem to be "getting at the same thing." If so, a name can be given to each cluster or factor so identified.

Quinn and his colleagues conducted such an analysis on the importance ratings of half the respondents in the 1969 Quality of Employment Survey. They then tested the results by repeating the process on the other half of the same sample and on a number of subgroups—men and women, black and white, old and young. The entire process was repeated in 1973, when the second Quality of Employment Survey was conducted. The results, which are extremely stable, are summarized in Table 3.1. The statements that make up the body of that table appear exactly as they did on the cards that survey respondents were asked to sort. The number following each statement is the percentage of people in the country as a whole who considered that statement to describe something *very important* in a job.

One of the common criticisms of factor analysis is that "you get out what you put in." In one sense the truth of the statement cannot be denied; the process of searching for factors or clusters of items is limited to those items that were included in the first place—limited, in this example, to the deck of cards that interviewers gave to respondents. But in another sense you do get out more than you put in; you discover the structure or patterning of items that reflects the way people think about work, and that was the purpose of the inquiry.

A number of such factor-analytic studies have been done, and several summaries are available (Robinson, Athanasiou, and Head, 1969; Smith, Kendall, and Hulin, 1969; Locke, 1976).

---

[1]For a more detailed description of this procedure, see Hunt, Schupp, and Cobb (1966).

Table 3.1   Factor Structure: Importance of Job Characteristics[a]

	Percentage Who Rate Statement Very Important
1. Task content (factor II: challenge)	
(a) The work is interesting	76
(b) I have an opportunity to develop my own special abilities	69
(c) I can see the results of my work	64
(d) I am given a chance to do the things I do best	59
(e) I am given a lot of freedom to decide how I do my own work	53
(f) The problems I am expected to solve are hard enough	25

2. Autonomy and control

The Quality of Employment Survey included few statements that were intended to measure autonomy, and none that attempted to measure the extent of worker influence or control over aspects of organizational life beyond the immediate job. Item 1(e) appears to be a direct statement about autonomy, but the factor analysis indicated that people thought of it as fitting with other challenging task characteristics rather than as separate from them.

3. Supervision and resources (factor V: resource adequacy)	
(a) I have enough information to get the job done	72
(b) I receive enough help and equipment to get the job done	69
(c) I have enough authority to do my job	68
(d) My supervisor is competent in doing his/her job	64
(e) My responsibilities are clearly defined	63
(f) The people I work with are competent in doing their jobs	59
(g) My supervisor is very concerned about the welfare of those under him/her	55
(h) My supervisor is successful in getting people to work together	54
(i) My supervisor is helpful to me in getting my job done	52
(j) The people I work with are helpful to me in getting my job done	49
(k) My supervisor is friendly	48

Table 3.1 (Continued)

		Percentage Who Rate Statement Very Important
4.	Relations with co-workers (factor IV)	
	(a) The people I work with are friendly and helpful[b]	70
	(b) The people I work with are friendly	54
	(c) I am given a lot of chances to make friends	40
	(d) The people I work with take a personal interest in me	31
5.	Wages (factor III: financial rewards)	
	(a) The pay is good	64
	(b) The job security is good	62
	(c) My fringe benefits are good	53
6.	Promotions (factor VI)	
	(a) Promotions are handled fairly	61
	(b) The chances for promotion are good	57
	(c) My employer is concerned about giving everyone a change to get ahead	54
7.	Working conditions (factor I: comfort)	
	(a) I have enough time to get the job done	53
	(b) The hours are good	46
	(c) Travel to and from work is convenient	44
	(d) The physical surroundings are pleasant	40
	(e) I can forget about my personal problems	27
	(f) I am free from the conflicting demands that other people make of me	26
	(g) I am not asked to do excessive amounts of work	18

[a]The results of the factor analysis are presented in the same order as the eight key characteristics discussed in the preceding section of this chapter. The labels of these characteristics are also used, but the factor number and name as proposed by Quinn and Cobb (1973) are in parentheses (Quinn and Shepard, 1974).
[b]This statement, which was asked in the 1969 survey, exemplifies a common error in question construction; it is "double barreled." Responses are therefore ambiguous, since we cannot know to what extent the person is responding to the term "friendly" and to what extent to the term "helpful." In the 1973 survey the item was repeated for the sake of comparability with the earlier study, but two items were added that separate the terms "friendly" and "helpful." They appear as items 3(f) and 4(b). The helpfulness of co-workers is apparently thought of more as a contribution to one's resources than as an expression of personal warmth and interest.

Among these studies, the Quality of Employment Surveys are distinctive in three respects: they are based on national probability samples of substantial size, they have been replicated three times within a decade, and they are based on individual statements about what is important in a job rather than on expressions of current job satisfaction.

The ratings of importance enter into the data of Table 3.1 in two ways. They are the basis for the factor structure that defines the numbered subdivisions of the table (task content, supervision, etc.), and they also indicate the relative importance that people attach to the components and the items themselves. In general, people rate the content of their jobs and the resources and rewards for doing them as more important than peer relations or working conditions. There is more variation in the importance ratings of items within any factor, however, than between the average ratings of the factors.

If we select those items that were rated as very important by more than 60 percent of employed men and women throughout the United States, we emerge with a profile description of a good job. "A good job," our composite respondent tells us, "is one in which the work is interesting, I have a chance to develop my own special abilities, and I can see the results of my work. It is a job where I have enough information, enough help and equipment, and enough authority to get the job done. It is a job where the supervisor is competent and my responsibilities are clearly defined. The people I work with are friendly and helpful; the pay is good, and so is the job security."

THE MICHIGAN ORGANIZATIONAL ASSESSMENT PACKAGE

Studies of work that depend entirely on the self-report of individual workers have taught us a great deal, but they have characteristic limitations as well as strengths. They show us the job through the eyes of the worker, and we therefore see those aspects of the work situation that the worker encounters directly.

The reported simplicity or complexity of the task, the rewards for its performance, the helpfulness of the supervisor, and the friendliness of co-workers have in common these properties of subjectivity and immediacy.

Except for the self-employed minority, however, work occurs in an organizational context; people work as members of organizations, and their jobs are affected by organizational properties. Size, number of hierarchical levels, growth rate, competitive position, managerial philosophy, technology, and other characteristics of organizations as such can be thought of as indirect determinants of the individual's work experience. They affect the individual mainly through their effects on job characteristics already described.

Indirect effects, however, are not necessarily weak, and the effects of organizational characterisitics on the individual's work life are very strong indeed. The wage levels that are possible in an organization, for example, must depend on its overall financial position, and the content of jobs must reflect the technology of the organization as a whole. The Quality of Employment Study was based on a nationwide sample of individuals and gathered little information at the organizational level. Had it done so, there is little doubt that organizational characteristics would have added one or more factors to the six that emerged from the factor analyses of that study.

Such factors appear prominently in the Michigan Organizational Assessment Package, an ambitious collection of items, scales, and procedures for observation and archival analysis. The Assessment Package collection of items, scales and instructions for observation, shares the broad aim of the Quality of Employment surveys—to measure the quality of work experience and the factors that determine it. It shares also the emphasis on quantitative measures and their standarization and validation. In other respects, however, the Michigan Organizational Assessment Package is different from and complementary to the Quality of Employment Surveys. It was developed and tested in organizational settings rather than on population samples; it uses observational methods as well as self-report; it includes technical and economic data about the organization, and other measures that go beyond the job and the immediate work group.

The Organizational Assessment Package is also very large. The self-report questionnaire, which is only one part of the package, includes some 350 items and 100 scales. These were the survivors of a universe of about 1500 items. A condensed or core questionnaire consists of 150 items that are intended for inclusion in every organizational study using the package; other measures are to be chosen from the package according to the special purposes of the inquiry and the characteristics of the organization under study.

To facilitate such choices, the Organization Assessment Package is divided into 10 modules. In most respects these modules are consistent with the factors indentified in the Quality of Employment Survey—task and role characteristics, work group functioning, supervisory behavior, compensation, and performance evaluation. They go beyond those factors in their detail and in their coverage of two areas—autonomy and control, and organizational context—included in the list of key occupational characteristics.

The Organizational Assessment Package includes a module called influence structure, which enlarges on the aspects of autonomy and control discussed earlier (key characteristic 2). Four main scales are included as measures of the employee's involvement in the influence structure of the organization—influence over one's own work activities, influence over the coordination of one's own work with that of others, influence over the assignment of other people to tasks, and general influence in organizational matters. These scales can be regarded as measuring influence over successively broader issues within the organization, beginning with one's own job and ending with matters of more general policy.

The package puts great emphasis on organizational context (key characteristic 8). Measurements of organizational performance include rates of absence and tardiness, turnover, accidents, grievances, and days lost because of strikes. Other economic measures at the organizational level are inventory shrinkage, machine repair, productivity, and product quality—all of which are reduced to dollar units. The technology of the organization is measured in terms of task cycle time, delay time, number of operations involved in completing the organizational product, and the like.

Even within the self-report part of the package, there is a module on intergroup relations, which includes issues that go beyond the immediate job of the individual. For example, there are scales on the required interdependence among work groups, the amount of conflict among interdependent groups, the ways such conflicts are resolved, and the resulting organizational climate.

DIFFERENTIATION OF JOB CHARACTERISTICS

The Michigan Organizational Assessment Package is important for its depth as well as for its range. For example, let us consider task content. The Quality of Employment Survey measures task content by means of six items that form a single factor; an inspection of the items suggests challenge as the dominant idea, and the factor was so named (Table 3.1). By contrast, the self-report questionnaire of the Michigan Organizational Assessment Package includes 60 items on the subject of task content, grouped into the following 16 scales:

1. Challenge—the extent to which the job involves one's skills and abilities.
2. Meaningfulness—importance and meaning of tasks to the individual.
3. Responsibility—importance of successful performance to the individual.
4. Variety and skill—number and complexity of tasks that make up the job.

   5. Task identity—production of an entire product or service.

   6. Task feedback—knowledge of one's performance from the work itself.

   7. Work influence—influence over decisions affecting one's own work.

   8. Autonomy—freedom to decide what and how things shall be done on one's job.

   9. Pace control—control over the speed at which one works.

 10. Role conflict—incompatible demands for performance of the job.

 11. Role clarity—knowledge of what is expected on the job.

 12. Task uncertainty—predictability of events and procedures for handling them.

 13. Task interdependence—extent to which job requires coordination with others.

 14. Role overload—performance requirements in relation to time constraints.

 15. Resource adequacy—availability of tools, supplies, and information.

 16. Skill adequacy—job demands in relation to skills and training of person.

These 16 scales represent a considerable elaboration of task content, primarily through a process of differentiation. For example, the idea of challenge to the person's interests and abilities, which is one aspect of task content, is expressed in at least the first six of the scales. The process of differentiation stops only when explanation is complete or fragments can no longer be divided. That point, as chemistry and physics have demonstrated to a baffling and impressive degree, is almost infinitely distant in scientific investigation. Each of the eight job characteristics discussed at the beginning of this chapter could be similarly differentiated. Even in their present form, however, they tell us much about what makes jobs good or bad.

SUMMARY

Widespread agreement about which jobs are good and which are bad leads to the question of what characteristics are reflected in these judgments. This question is important not only for scientific purposes but for a highly pragmatic reason as well; to improve the quality of jobs we must understand what qualities job-holders value. No single paradigm answers this question to the satisfaction of all investigators, but eight components of jobs are widely recognized as important determinants of their overall quality: task content, autonomy and control, supervision and resources, relations with co-workers, wages, promotions, working conditions, and organizational context. Measures of these job components and research findings about them are presented from various sources, especially the nationwide Quality of Employment Surveys and the Michigan Organizational Assessment Package.

CHAPTER FOUR

Stress and Strain: The Evidence

Stress and strain are old words, from the Latin and Anglo-Saxon, and they have been in the English language for a long time. Their paired meaning to describe situations that cause discomfort or pain, however, is a recent borrowing from physics and engineering. In those fields both terms have unambiguous definitions: Stress is a force or pressure applied to some object; strain is the resulting change in shape or volume of the stressed object. The force of the wind against a building and the weight of traffic on a bridge are stresses; the resulting sway and sag are strains.

The job characteristics reviewed in Chapter 3 are dimensions along which stress can be measured. Increases in work load, for example, can be thought of as increments of stress, in much the same way that we might think of increasing weight on a bridge or building. The "engineering analogy" (Lazarus, 1966) is attractive; it makes job stresses seem more understandable and more measurable.

A great deal of stress research, especially in the laboratory, has been done in terms of the engineering model. The researcher imposes some unpleasant stimulus and then observes the effects. Let us consider some examples of such laboratory research, chosen for their seeming relevance to work situations.

54

STRESS IN THE LABORATORY

Task Content

The physical demand of experimental tasks produces physiological signs of strain (elevated heart rate, increased secretion of adrenaline and noradrenaline, elevated systolic blood pressure, etc.) only if the physical demand is heavy or if it is perceived as heavy (Frankenhaeuser, 1971). Mental tasks (e.g., mental arithmetic) have also been shown to evoke physiological strain, at least when the tasks are performed under distracting conditions. These have included recorded workshop noise (Frankenhaeuser, 1971), flickering light (Raab, 1966), and other irritating factors of the kind presumably encountered at work (Frankenhaeuser and Patkai, 1964). All these experiments involved characteristically brief periods of task performance. Levi (1972), however, conducted an experiment in which subjects (army officers and corporals) engaged in a simulated electronic firing of rifles against moving targets for a period of 72 hours. There was a marked decrement in performance and a marked increase in physiological indicators of strain (adrenaline and protein-bound iodine), together with considerable subjective distress.

Physiological signs of strain have been evoked experimentally by embarrassing interrogation (Hamburg, 1962), by the anticipation of uncomfortable or annoying situations (Berman and Goodall, 1960), and by the anticipation of tasks characterized by uncertainty of success or penalty (Frankenhaeuser 1971).

Physical Conditions

Much of the research on physical conditions as stressors has been done on animals, and the extrapolation to human beings in general and the work role in particular remains uncertain. Nevertheless, it is worth remembering that acute increases in blood pressure and other symptoms of strain have been produced in laboratory animals by means of long-lasting noise, crowding, conflicting messages of stimulation and inhibition, and physical bumps or blows (Brod, 1971).

A few of the animal experiments on stress seem especially relevant to the stresses of work. One series dealt with the differences between stresses taken singly and in combination (Dean, 1966). Rats were subjected separately to extreme heat (115°F), simulated high altitude, and random vibration, and also to combinations of these conditions. Animal deaths under the separate conditions varied from 0 to 7.5 percent, but heat and vibration in combination caused a 65 percent death rate, and altitude and vibration had similar results. The significance of the experiment lies in the fact that work situations often produce stresses in combination, while the canons of experimental research are usually interpreted to mean that stimuli should be studied one at a time.

Another line of animal experimentation that seems important for understanding the work role involves the imposition of punishment (electric shock) under various task-relevant circumstances, in an effort to discover to what extent the punishment itself is the source of strain and to what extent the task-related circumstances of its imposition. Some findings (Corson, 1971) suggest that the combination of unavoidable stress, insoluble problems, and inescapable punishment causes the most strain (increased heart rate, respiration rate, and temperature in dogs). Other experiments, including those of the widely publicized "executive monkeys," seem to argue that strain is greatest when the punishment of two or more animals depends on the performance of one of them.

The Stress of Responsibility

In these experiments Rioch (1971) altered the usual procedures of stress research with animals in a way that seems to capture the particular stress of responsibility for the well-being of others. His experiments involved pairs of monkeys, visible to each other, engaged in a recurring six-hour "task" situation. During the "work" period, a shock was scheduled for application at regular intervals, which were sometimes as brief as 20 seconds. Each monkey had access to a lever, and one of the levers was "preventive." If the preventive lever was pressed, the shock was avoided, not only for the monkey that pressed it but for the other monkey as well; hence the term "executive" was used for the animal whose behavior permitted or prevented the punishing shock. The design of the experiment ensured that the exposure of both animals to the electric shock would be identical; they escaped or were punished together, but only one of them—the executive— could prevent the shock. All the "executive" monkeys died of gastrointestinal lesions in periods varying from 9 to 48 days; the nonexecutive monkeys received the same shocks, but none died and none showed gastrointestinal pathology.

Too Much and Too Little

Most laboratory experiments create stress by imposing too much of something—noise, physical demand, responsibility, and the like. Such stresses, consistent with the engineering analogy, have been shown to cause strain. But living creatures, unlike bridges and buildings, show signs of strain when there is too little demand on their capacities as well as when there is too much.

Hebb (1958) conducted an experiment, for example, in which the participants were required to do literally nothing. People rated the situation "unbearable" after three or four days. Within that period they showed tension, sleeplessness, personality changes, reduced intellectual performance, and feelings of "depersonalization"; all these symptoms quickly disappeared when the subjects were no longer isolated and inactive. The symptoms were evoked by experimental conditions that involved no restraint of movement except for remaining in a small isolation room.

Frankenhaeuser (1971) reported increases of adrenaline and noradrenaline under more moderate conditions of understimulation—the requirement to press a button in response to an irregular signal under conditions of isolation from others. The signs of strain increased when the task was made more complex—the matching of responses by buttons and pedals to multiple signals. This combination of specialized overload and general understimulation (lack of opportunity for interaction with others, limitation of activities to a small but demanding repertoire) is suggestive of many work roles, white collar and blue collar, in advanced technologies.

A few experiments have attempted to simulate such work roles under realistic conditions. Levi (1972) had groups of subjects sort ball bearings according to small differences in size. The task was monotonous and demanded neither much learning nor much skill. But it did require effort and attention; mistakes were easy to make, and there were time pressures to complete the job, critical observations about the performance, variations in light, and a background of recorded workshop noise. The task required about two hours of such work. It produced subjective ratings of unpleasantness and physiological changes of increased heart rate and blood pressure, increased secretion of adrenaline and noradrenaline, and increased concentrations of triglycerides and free fatty acids in the blood plasma. The experimenter comments that "it is tempting to speculate about the effects of the socio-economic or other real-life stressors, which may be repeated over months and years and surely may represent a threat to the individual far exceeding that implied in our laboratory situation" (Levi, 1972, pp. 91–105).

STRESS AND STRAIN: FIELD STUDIES

Studies of work stress have been done in real-life situations. Such studies have the potential advantages of large numbers, representative samples, and the sustained character of the situation under observation. They also have the power and realism that comes when people are "playing for keeps" rather than assuming the brief role of experimental subjects.

With these advantages go the characteristic disadvantages of nonexperimental research in real-life settings. The stresses of life are not arranged for the convenience of the researcher. They are not nicely controlled or neatly partitioned. They come in complex combinations. The people who endure them are not randomly chosen, and the stresses are encountered under a bewildering variety of conditions.

The problem in such research is not to discover signs of strain; they are unhappily abundant. The problem is to identify the causes, to discover the stressors, and separate them from the tangle of surrounding data. Some researchers have approached this problem by making comparisons *within* an occupational group, on the assumption that people engaged in the same occupation are similar in many respects and that some stresses differ significantly within occupations.

For example, Russek (1962) studied certain symptoms of strain within three professions—medicine, dentistry, and law. Within each of these occupational groups, he correctly predicted the relative frequency of hypertension and coronary heart disease. For medicine, the incidence of these diseases increased from pathologists and dermatologists to anesthesiologists and then to general practitioners. For dentistry the incidence of these diseases was successively higher among periodontists, orthodontists, oral surgeons, and general practitioners. For law, the ordering was patent law, other specialties, trial law, and general practice. The inferred stressor in this research seems to be direct responsibility for the well-being of others, especially in combination with the necessity of being directly responsive to many people under continuous time pressure.

Time Pressure and Responsibility

The combined stressfulness of responsibility and time pressure is implied in several other studies in which comparisons were made within either occupational or industrial groups.

According to Doll and Jones (1951), foremen and executives in

England have more ulcer disease than the people they supervise; Vertin (1954) found that foremen and assistant foremen in Holland have more peptic ulcer than workers under their supervision. Similar findings are reported elsewhere in Europe (Pflanz, Rosenstein, and Von Uexkull, 1956; Gosling, 1958) and in the United States (Cobb, 1974). In one large data set (4325 air traffic controllers compared to 8435 second-class airmen), the inference of responsibility is particularly compelling. Hypertension is four times more common among the air traffic controllers, and diabetes and peptic ulcer more than twice as common. Moreover, all three diseases show earlier dates of onset, and two of them (hypertension and peptic ulcer) show a "traffic effect"—that is, they are more common among traffic controllers assigned to the busier airports (Cobb, 1974; Rose, et al., 1978).

Overload and Underutilization

The stresses of responsibility for others are very real and they have real effects. They are not, however, the most prevalent of job stresses. Many jobs present problems of qualitative underutilization in combination with quantitative overload; they use few of the worker's skills and abilities but they make heavy demands on those few. This stress pattern and its effects are shown clearly in an intensive field study conducted in Swedish sawmills and lumber-trimming plants (Gardell, 1976).

Three categories of jobs were selected for study, a "risk" group and two control or comparison groups. The risk group consisted of jobs with extremely short operating cycles (usually less than 10 seconds), a total dependence on machines (saws, edgers, trimmers, etc.) for determining the work pace, sustained positional and postural constraints, and a need for unremitting attentiveness. The first comparison group included men who worked in the same factories but were not subjected to the same control by machines; their jobs also required less skill than those in the risk group. The second comparison group consisted of maintenance workers, whose jobs are high in skill and autonomy. Other fac-

across jobs—age, sex, seniority, and (with
work and methods of pay.

a sharp and consistent pattern of perceived
t, and symptoms of strain in the risk group
compared to the other two groups. Consistent with this pattern
is the higher rate of absence, both general and stress-attributed.
Workers in the risk group also showed lower patterns of off-the-
job participation in organizations, and even their participation
in union activity was low. These findings are summarized in Table 4.1.

**Table 4.1   Job Comparisons in the Sawmill Industry
(Sweden): Responses to Objective Stress (Montony, Lack
of Autonomy)[a]**

	Risk Group (%)	Control Group (%) I	II
Workers' perceptions of job			
Very monotonous	43	0	8
Hectic work pace	36	10	0
Work prevents social contact	43	0	0
Workers' affective responses			
Loath to go to work	50	0	0
Physically weary after work	36	10	0
Mentally weary after work	43	10	0
Worried about health risks	29	10	0
Physiological responses			
Elevated blood pressure	14	10	0
Headache	36	0	20
Nervous complaints	36	0	0
Gastric ulcer and catarrh	50	50	30
Back ailments	43	10	20
Behavioral responses			
Malaise attributed to work	57	20	25
30 or more days absent[b]	29	0	0
One or more days absent (stress)[b]	43	0	0

[a]From B. Gardell, *Job Content and Quality of Life* (Stockholm: Prisma, 1976).
[b]Absences, for all causes combined and for reasons of work stress, refer to the
twelve months preceding the research.

Physiological stress reactions were also measured for men in these groups, by means of the amount of adrenaline and noradrenaline in the urine. Results of this analysis are presented in Figure 4.1, which shows significant differences between the risk group and the combined control group, both in the level and in the pattern of adrenaline output during the workday.

The secretion of adrenaline and noradrenaline has been used extensively in laboratory research to measure physiological responses to a variety of stimuli regarded as stressful (Frankenhaeuser, 1971), and its reliability and validity are well established. In the research cited here, a base value was computed for each individual while "at rest." Readings at four times during the workday are shown averaged as percentages of these base

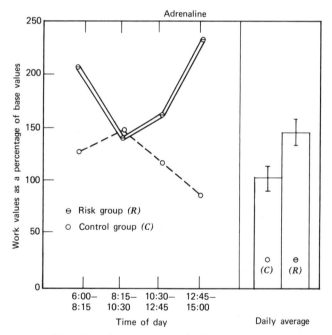

Figure 4.1. Physiological stress as measured by excretion of adrenaline and noradrenaline [from B. Gardell, *Job Content and Quality of Life* (Stockholm: Prisma, 1976)].

values, and the bar charts represent the group averages for the workday.

The overall stress effect of the workday is thus shown by the bars, which indicate that the risk group has a secretion rate during the workday that is about 150 percent of its rate at rest, whereas the average secretion rate of the control group at work is little higher than its rate at rest. The pattern of difference during the workday is even more impressive and is congruent with the subjective reports of the two groups.

The control group begins the day with a secretion rate only a little above that at rest; the secretion peaks early and "winds down" gradually to an end-of-day reading slightly below rest level. By contrast, workers in the risk group show considerable anticipatory stress at the beginning of the workday and, after a mid-morning reduction to rest level, become increasingly stressed, so that their peak reading (more than twice their own base level) occurs at the very end of the workday. This pattern of catecholamine secretion fits the workers' statements of reluctance and anxiety about going to work and their avoidance of social activities after work. As one of them said, "It takes at least a few hours to get all that work rhythm and noise out of your system and before you even feel up to . . . [being] with the family" (Gardell, 1976).

A MODEL FOR STRESS RESEARCH

A number of models have been proposed for the study of stress, either to replace or to elaborate the simple proposition that stress causes strain. One such model (French and Kahn, 1962) was developed to provide a framework for research on work and health, and it serves that purpose for a continuing series of investigations at the Institute for Social Research (ISR) of the University of Michigan.

The core of the model (Figure 4.2) is a causal sequence that leads from characteristics of the objective work environment to

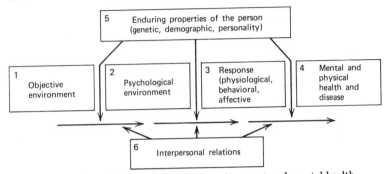

Figure 4.2. ISR model: the social environment and mental health.

the worker's psychological or subjective experience of these characteristics, his or her immediate response, and the longer-term effects on mental and physical health.

The ISR model is extremely broad when presented in this way; it is in fact a set of conceptual categories rather than the representation of a theory. It becomes both more restrictive and more informative as specific concepts are introduced into the several categories (boxes) and connections (arrows) are proposed between them. Thus connections of the A→B category have to do with the effects of the objective work environment on the psychological work environment (the work environment as the individual experiences it). For example, salespersons, union stewards, and other people whose jobs require them to engage in transactions across the organizational boundary (a fact in their objective work environment) more often report that they are subjected to incompatible demands on the job (a fact in their psychological environment).

Connections of the B→C category relate facts in the psychological environment to the immediate responses that are invoked in the person. For example, the perception that one is subject to persistent conflicting demands on the job is associated with feelings of tension. The C→D category deals with the effect of such responses on criteria of health and illness. The relationship of sustained job tension to coronary heart disease (perhaps more arguable with respect to the empirical evidence) illustrates the $C \rightarrow D$ category.

Finally, the three categories of connections just described must be qualified by an additional class, represented by the vertical arrows in Figure 4.2. This class states that relationships between the objective and psychological environment, between the psychological environment and the response, and between the response and health or illness are modified by enduring properties of the individual and by interpersonal relations.

For example, the extent to which a person experiences tension on being exposed to role conflict depends very much on the personality characteristics of flexibility—rigidity; people who are flexible rather than rigid respond with greater tension to the experience of role conflict. Supportive relations with others, however, seem to buffer or modify some of the relationships between the stressful demands of the work role and the consequences for the individual. French (1974) found in a study of a government agency that quantitative work load was related to diastolic blood pressure ($r = .33$), but not among those employees who had supportive relationships with their supervisors. In similar fashion other properties of the person and his or her interpersonal relations act as conditioning or modifying variables in the causal sequence that leads from the work role to health or illness (La Rocco, House, and French, 1980).

Given this framework, we can define an adequate explanatory sequence; it would consist of a chain of hypotheses beginning with some characteristic of the objective work environment, ending with some criterion of health, specifying the intervening variables in the psychological environment and in the immediate responses of the individual, and stating the ways in which this causal linkage is modified by the differing characteristics of individuals and their interpersonal relations. Such a causal chain might be called a theme, and a set of such themes, logically related, would constitute a theory of work and health.

No one could claim that such a matured theory yet exists, much less that its components have been subjected to empirical tests. The framework serves to remind us, however, of one form that such a theory might take. It reminds us also of the cumulative meaning of such empirical fragments as are presently available to us regarding the social and psychological aspects of work and health.

SUMMARY

Studies of stress have become frequent, both in the laboratory and in the field, and some of the job characteristics discussed in earlier chapters now appear as sources of stress. Laboratory studies have measured the physiological effects of heavy task demands, both physical and mental. Those effects include elevated heart rate and systolic blood pressure and increased secretion of adrenaline and noradrenaline. Heat, vibration, and simulated high altitude also have damaging effects, especially when they are imposed in combination. Understimulation has been shown to be stressful, in much the same way as overload, and one series of studies suggests that responsibility for the welfare of others in addition to oneself is highly stressful.

The results of field studies in work settings are generally compatible with findings from the laboratory. Responsibility, time pressure, overload, and underutilization affect physical symptoms and catecholamine secretion as well as satisfaction and dissatisfaction. A model for the study of such stresses is proposed.

CHAPTER FIVE

Unemployment: The Greater Stress

When people are asked to tell about times of joy and satisfaction in their lives, they talk of their husbands or wives, their sons and daughters, their work, and their friends. But when they are asked about the problems and dissatisfactions in their lives, they speak again of family, work, and other close attachments. Each of us, I suspect, can testify to the validity of these responses. The important relationships and activities in our lives are sources both of our greatest satisfactions and our greatest problems.

It follows that we should be concerned with the improvement of these attachments rather than with their dissolution. To put the case in the context of work, being dissatisfied with one's job does not imply that one would be more satisfied without it. The discovery that some jobs are unreasonably stressful calls for the reduction of stress, not the removal of work. A job must be very bad indeed before it is worse than none at all.

People understand all this, which is why most of them say, as we have seen, that they would keep working even if money were not a consideration. The sense of activity, of doing something that others value and require, and of joining with others so engaged at a particular time and place are important sources of meaning in life. In principle, activities other than paid employ-

67

ment can meet these criteria, but for most people such alternatives are either unimagined or inaccessible.

THE PASSIONATE AMBIVALENCE

People are thus of two minds about work, aware of both its gratifications and frustrations, and many are attracted and repelled at the same time. Psychologists call this state of opposite feelings—which may be weak or strong—ambivalence.

Feelings about work run strong and deep. In 1960 Robert Weiss and I asked a national sample of employed men a direct question: What makes the difference between something you would call work and something you would not call work? Almost half the people interviewed answered in terms of enjoyment: If an activity was enjoyable in itself, it was not work. This answer was more characteristic of factory and service workers than professionals and managers, who were more likely to define work as something necessary or externally required. Many people at all occupational levels distinguished between work and nonwork mainly in terms of pay. In short, work was seen as an activity not enjoyable in itself, but required and scheduled by external circumstances and persons, and (reasonably enough, therefore) it was an activity paid for by those who required it.

Ten years later, the same question was asked of a small sample of workers interviewed in connection with the Quality of Employment Survey (Chapter 3). There had been little change in those 10 years. Forty-five percent of the workers interviewed defined work as something not enjoyed. The notions of payment and requiredness as defining characteristics of work, although not mentioned quite as frequently as in the earlier study, still ranked high, as did the idea of work as exertion or activity.

And yet many of the same people who defined work in this fashion also said that they would go on working even if they didn't need the money. About three-quarters of all employed men and three-fifths of all employed women gave that answer in

1969 and 1973. Hence the judgment of ambivalence; workers are saying in effect: Work is an activity that is not enjoyable in itself, is scheduled and required of me by others, and for which I am paid. Yet I would continue to work even if I had no need for the pay.

I interpret these data to mean that for most men and for increasing numbers of women there is no viable alternative to work, no other activity that uses energy, demands attention, provides regular social interaction around some visible outcome on which the larger society confers some dollar value. Other findings bear out this interpretation. Two-thirds of the people who say they would go on working explain that work keeps them from being bored or gives some direction to their lives. And when those who say they would quit work if they could afford to are asked what they would miss, more than anything else they speak of their co-workers.

Critics of this work-oriented view argue that it is more descriptive of the past than the future. The argument takes two main forms, which we may designate the technological and the generational. The technological argument asserts that technology has changed the nature of work and that the work ethic is a casualty of those technological changes. Work that is either fractionated or automated beyond the point of human interest, so the argument goes, has created a work force without interest in work. Work has become less involving and workers have become less involved. The generational variant of this argument also assumes the demise of the work ethic but finds the explanation in the affluent, leisure-saturated experience of American youth. The work ethic, according to this version, is alive but not well; it is aging.

I believe that there is a little truth in the first of these arguments and almost none in the second. If the erosion of the work ethic were a generational phenomenon, we should find that more old people would choose to go on working, rich or poor, while more young ones would quit if they could afford to. The data show exactly the opposite; the younger the age group, the larger the proportion of those who would continue to work (See Figure 5.1.) It appears to be the fatigue of age rather than the disaffection of youth that accounts for the decision to quit work when it becomes financially possible.

If the alleged antiwork malaise were general rather than age-specific, each of the age curves in Figure 5.1 would be plunging sharply toward the zero line at which everybody quits work as early in life as possible. The data are by no means that dramatic. There was some downward slope between 1953 and 1969, and if it had continued, the day would have come when only the financial incentive would keep people on the job. Between 1969 and 1977, however, the data show a stable or slightly increasing tendency to choose work over nonwork.

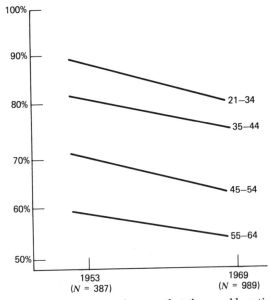

Figure 5.1. Percentage of men who state that they would continue working even if they had no economic need to do so, by age, for 1953 and 1969.

ATTACHMENT TO WORK

When workers say they would work even if they were independently wealthy, they are indicating the strength of their attachment to work. When they explain the reasons for their feelings or describe the things they would miss if they were not working, they are indicating the nature of the attachment—the aspects or features of work that they find attractive. The research of Robert Dubin and his colleagues (1976) is unusual in its exploration of both these issues. Their research, which includes some 18 studies conducted in five countries over a 20-year period, addresses two main questions:

1. To what features of work do workers become attached?
2. How strong are those attachments?

Dubin and his colleagues found that workers become attached to many features of work and that the bonds of attachment are strong. Moreover, both the number and the strength of these ties have been underestimated, a finding that has many implications for social theory and social policy, among them the failure to understand the full significance of unemployment. Most theories of work and work motivation have emphasized some particular form of work attachment and neglected or denied the importance of others.

For example, Taylor (1911), and the later practitioners of that form of time study and industrial engineering to which he gave the prestigious and undeserved title of scientific management, emphasized financial reward as the main attachment to work and gave little or no consideration to other bonds. The worker is regarded as an economic man (women workers are not discussed), seeking solely to maximize income within the limits of his physical capacity and the opportunities available to him.

The discovery, or rediscovery, of the work group as an object of attachment and gratification was something of an antidote to the "money only" theories of work. The importance of group membership and the consequent power of the group over the individual dominated a great deal of social research and application. The behavior of soldiers in battle was explained in this way (Stouffer, et al., 1949). The Hawthorne studies at Western Electric, best known for the alleged effect of experimental participation as a stimulus to productivity, are famous also for demonstrating the power of work groups to control the productivity of their members (Roethlisberger and Dickson, 1939). Studies by Lewin and his colleagues (Cartwright and Zander, 1960) demonstrated the power of the group and of participation in group decisions in many settings and circumstances—among them the adoption of new food habits by housewives during wartime shortages, the acceptance of changes in work requirements to meet external competition, and the regular conduct of self-examination for breast cancer.

These were sophisticated experiments in group dynamics. In their vulgar form, however, group-oriented explanations of why people work became almost as monolithic as the Taylorism they opposed. The affiliation motive replaced the drive for income maximization, and the social model of the human being replaced the economic one. The emphasis on friendly and supportive behavior by supervisors, which was sometimes identified as the human relations movement, can be considered another version of the affiliative model of human behavior, in which the relationship of the worker to the supervisor or formal leader was made the primary attachment and the main explanation of worker behavior.

More recent theories of work give some recognition to the multiplicity of human motives and their expression on the job. Likert's (1967) managerial theory, for example, rests almost equally on the three principles of group membership, participation in decision making, and supportive supervisory behavior. Maslow's (1954) motive hierarchy assumes that human beings have many needs and that they seek to gratify them in an ascending order of complexity and abstraction. As the biological requirements for food, water, and the like, are satisfied, needs for security and community become dominant; as these are satisfied, needs for self-esteem and self-actualization begin to dominate the individual's behavior. The more primitive needs do not vanish, of course, and they quickly reassert their dominance if they are not satisfied. The implication is that different job characteristics are required to satisfy these various needs and that attachments of a person to a job can develop around any of them, but the connections are not made specific.

Katz and Kahn (1978) distinguish three main motivational patterns at work: compliance with rules, responsiveness to rewards, and internalized motivation. All three are assumed to be active to some extent in most settings and for most people, and each involves distinctive attachments to work. In compliance with rules, the main attachments are to authority figures and to the concepts of obedience to law and legitimate authority. In the reward-motivated pattern of work the attachments are to pay, promotion, praise, and recognition. And in the pattern of internalized motivation the attachment is to the task itself, the group with which it is accomplished, or to the organization itself and the values it embodies.

As theories of human motivation and human organizations become more comprehensive, it becomes apparent that the number of possible work attachments is very large. It is a strength of the Dubin research (1976) that he and his colleagues have attempted to provide a framework for these attachments, have developed ways of measuring them, and have applied those measures to large and diverse populations of workers. Their classification of work attachments begins with the distinction among three features of the work environment: the *human systems* (individuals and sets of individuals) that operate within it, the *objects and conditions* that exist in it, and the *payoffs* that it provides. Within these three main categories, the specific features to which attachments might be formed are then enumerated (Table 5.1).

Table 5.1   Attachments to Work[a]

Systems of the work environment
  Self
  Work group
  Company
  Union
  Craft or profession
  Industry

Work place objects and human conditions
  Technology
  Product
  Routine
  Autonomy
  Personal space and things

Payoffs
  Money
  Perquisites
  Power
  Authority
  Status
  Career

[a]From R. Dubin, R.A. Hedley and T.C. Taveggia, Attachment to work, in R. Dubin (Ed.), *Handbook of Work, Organization, and Society* (Chicago: Rand McNally, 1976), p. 290.

Many of these are obvious in their potential for attachment, and the attachment is readily observable. The affection of a worker for his or her work group, the care and concern expressed for one's tools or machine, the defensive possession of one's office or space, and the careful comparison of one's wages and perquisites to those of others are familiar to everyone who has worked in an organizational setting or observed those who do. When the attachments are to more abstract features of work—status or autonomy, for example—the evidence is less direct but no less compelling.

Dubin and his colleagues measured these forms of work attachment in a very straightforward way. Participants in their research were presented with a list of 94 brief phrases or statements, each of which described some feature of the work environment. The workers were then asked to indicate all the statements that described some feature they considered important in their own work. This procedure was followed for large numbers of workers in many organizational settings, engaged in many different kinds of work, and in a number of different countries.

Perhaps the most important finding is the multiplicity of work attachments, the variety of ways in which people are attached to their work. Some general tendencies are apparent, however. In 14 studies for which comparative data are available, the proportion of workers expressing attachment to the formal and organizational aspects of their jobs ranged (with one exception) from 62 to 96 percent. Attachments to technical aspects of the job ranged from 54 to 90 percent. The range for reported attachments to informal aspects of the work situation was much lower, 9 to 52 percent.

The number and nature of attachments vary for different people and different work settings. For example, young workers are more likely than older ones to mention the work group, the technology, the prospects for early promotion, and the potential for a career as points of attachment to the job. Long-service employees give higher ratings than newer employees to the craft or profession, the work routine itself, and to their personal space or things. Women are more likely than men to mention their attachment to the work group, personal space or things, the product, and the company.

More interesting than these comparisons are the similarities and differences between people who consider their central life interest to lie in the domain of work and those who do not. This distribution was made by presenting participants with a list of activities that were equally likely to occur in work and nonwork settings and asking them to indicate the setting in which they would prefer to engage in each activity. One can, for example, help someone who needs it, talk to a friend, solve a problem, use a valued skill or ability, win an argument, and enjoy the recognition of others on the job or away from it. People who preferred the work setting as the place for these activities were said to have work as their central life interest.

The differences in work attachment between these two groups are consistent and plausible. People for whom work is the primary interest give more importance to such attachment items as "my supervisor's confidence in me," "my responsibility in my job," "the usefulness of the products or services created by the organization," and "chances for advancement and promotion." People who prefer to pursue their interests off the job were more likely to mention the following kinds of job characteristics as points of attachment: "having time for my personal needs," "knowing tasks in advance each day," "talking to others while working," "methods of payment" (hourly, piece rates, etc.), and "how modern the firm is."

For this discussion, however, the convergences in these data are more important than the differences. The attachments to work are many and strong; that fact holds for old and young, long-service and short-term workers, men and women, those who live to work and those who work to live.

REDISCOVERING THE IMPORTANCE OF WORK

Systems analysts tell us that the crucial test of what any part of a system contributes is to remove it. If you are in doubt about what a spark plug does for the engine, take it out and see how the engine runs without it. I believe that the ethical principles of social scientists are too strong, and their social power too weak, to permit experiments in which people are deprived of work. And such experiments are not necessary; business fluctuations, technological changes, plant relocations, and other such events make job loss and unemployment common during prosperous times and epidemic when times are bad. "Natural" experiments on the effects of unemployment, unfortunately, are always possible.

Studies of unemployment tend to be launched when rates of unemployment are high, and the results of such studies over a period of almost half a century show a familiar pattern. Especially for men and especially during the "working years" as society defines them, unemployment is a stressful state. In times past, the economic privation that accompanied job loss was so severe that the additional effects of other stresses were obliterated or ignored. In more recent years, when unemployment compensation and other benefits buffer many workers against the worst economic effects of job loss, the noneconomic effects of work deprivation have become more apparent.

They are destructive. As a result, people who do not have jobs try to find employment. People who do have jobs hold on to them until old age or ill health makes work more difficult and not working more acceptable. Those factors, along with eligibility for pensions and Social Security benefits, are the chief elements in retirement decisions. For people who are forced out of work for extended periods or prevented from getting it in the first place, the damage of unemployment is very great. It is ironic that these obvious facts should have to be so often rediscovered. That they must perhaps reflects the societal reluctance to recognize and correct them rather than any ambiguity in the facts themselves.

SOME STUDIES OF UNEMPLOYMENT

An early and classic study of unemployment was done in the Austrian town of Marienthal in the early years of the European depression (Lazarsfeld-Jahoda and Zeisel, 1933). It included careful observations and descriptions of the daily activities of unemployed men and showed a progressive reduction of their personal involvements and even of their physical movements.

During the depression years in the United States, similar findings of reduced activity in stricken towns and industrial areas were reported in studies conducted by the Work Projects Administration, a federal agency created to cope with unemployment by providing jobs, from building roads to painting pictures.

Other studies of the same period (Komarovsky, 1940; Bakke, 1940) reported a process of disengagement among the unemployed. The need to reduce expenditures forces unemployed workers to give up some forms of recreation and participation, of course. But the humiliation of being without work, the resentment that the "helping" organizations are not helpful enough, and the feeling that friends are being disloyal add to the tendency to withdraw. Withdrawal in turn deepens the feelings that caused it, and thus the cycle continues.

Most research on unemployment concentrates on some episode in people's lives or some moment in the life of a community. Much less is known about a person's lifetime experience with unemployment, although it seems certain that for some people unemployment is a recurring problem, and that recurrence adds to the damaging effects of being without a job. Wilensky (1961) investigated this issue by comparing men with "orderly" careers to those whose careers were "disorderly"—that is, marked by recurring instability of employment. Men with orderly careers had more frequent contact with relatives and friends, co-workers and neighbors. Moreover, these relationships were more likely to be experienced as compatible and to involve groups and organizations in the larger community. I interpret these findings to mean that reasonable stability of employment, predictability of one's working future, and a sense of "orderly progression along career paths" are important for forming and maintaining social relationships. Without long-term work stability and perspective, the worker tends to become isolated and less able to cope with either the unemployment problem itself or with its many derivative difficulties.

PLANT CLOSINGS IN DETROIT

Detroit may have the unhappy distinction of being a laboratory for research on unemployment. The automobile industry, still dominant in Detroit and other areas of southeastern Michigan, is notoriously unstable, and its instability extends to the smaller businesses that feed it and to those that meet the consumer needs of its workers. Its present difficulties repeat an unhappy and long familiar pattern. During the recession year of 1958, for example, the unemployment rate among blue-collar workers was 9.5 percent nationally and 33 percent in Detroit (Cohen, Haber, and Mueller, 1960). The high rate of unemployment lasted beyond 1958 in Detroit, and its impact on living standards was somewhat severe, as indicated by the measures that unemployed families reported taking. The adjustments were drastic, however, for unemployed families throughout the country: 42 percent withdrew money from savings; 44 percent postponed usual purchases; 30 percent left bills unpaid; 23 percent borrowed money and an equal number sought help from relatives; 13 percent moved to cheaper housing, and 10 percent went on relief (Cohen, Haber, and Mueller, 1960).

A detailed study was made of a long-time Detroit firm that closed during that recession period—the Packard Motor Company, one of the earliest producers of luxury automobiles. The closing of the Packard plant, after almost 60 years of continuous operation, produced among its workers a classic pattern of denial, resentment, and bereavement. The following quotations are taken from the report to the U.S. Senate Special Committee on Unemployment Problems (Sheppard, Ferman, and Faber, 1959):

I couldn't believe it [the rumors for two years].

I just never believed it would shut down.

I felt like someone had hit me with a sledge hammer.

I felt like a bomb hit me . . . no place to go.

It was a rotten thing to do . . . but everybody looks out for his own interests.

Well, I thought it was a dirty rotten deal. I could say some words you wouldn't want to write down about that.

Pretty damned mad . . . burned up . . . worked up inside.

I felt way down in the mouth . . . depressed.

A heartbroken business—just like if my wife died . . . when I went through that plant and saw those lines where cars came off, tears came to my eyes.

I could have cried. It's like losing your home.

One man committed suicide, he was behind in the rent on a new home.

I felt like jumping off Belle Isle bridge in the river . . . put that down, you put that down.

Most of the Packard workers (78 percent) eventually found jobs, but almost half of these who did find jobs were unemployed again at least once within two years after the original plant closing. Both the duration of the initial unemployment and the instability of the new job varied with the worker's age and level of skill. Longterm unemployment (more than one year) increased steadily with age, from 21 percent among those under 45 years old to 87 percent among those 65 or older. The relative skill of the worker had an effect almost as strong: 10 percent of the skilled and 28 percent of the unskilled were unemployed for more than a year.

Little is known about the effects of these experiences on the physical and mental health of the people involved, but the bits of information available suggest feelings of distrust and pessimism, a growing concern for job security, and a conviction that the government should somehow provide that security.

The judgment of pessimism and distrust is based on responses to the following eight statements, which comprise an index of anomie (Sheppard, Ferman, and Faber, 1959):

1. It is hardly fair to bring a child into the world the way things look now.

2. Most people don't really care what happens to the next fellow.

3. These days I get the feeling that I'm just not part of things.

4. You sometimes can't help wondering whether life is worthwhile any more.

5. In spite of what some people say, the lot of the average man is getting worse, not better.

6. These days I find myself giving up hope of trying to improve myself.

7. To make money there are no right or wrong ways any more, only easy ways and hard ways.

8. Nowadays a person has to live pretty much for today and let tomorrow take care of itself.

Most of the unemployed Packard workers agreed with most of those statements, and the proportion of agreement increased with the duration of unemployment (Table 5.2).

Table 5.2   Anomie, by Length of Unemployment, 1958 Sample[a]

Degree of Anomie[b]	Length of Unemployment		
	One Month or Less	One to Six Months	More than Six Months
Low	55	35	33
Medium	35	48	42
High	10	17	25
Total	100	100	100
Number of cases	31	80	171

[a]From H.L. Sheppard, L.A. Ferman, and S. Faber, Too old to work—too young to retire: A case study of a permanent plant shutdown, U.S. Senate Special Committee on Unemployment Problems (Washington, D.C.: U.S. Government Printing Office, 1959).

[b]Low, agreement with none to three statements; medium, agreement with four to five statements; high, agreement with six or more statements.

During the same period one of the three major newspapers in Detroit closed. *The Detroit Times* had been a high-circulation paper since the days of Hearst and Brisbane, with peak daily sales of more than half a million copies. Circulation began to drop, however, and the Hearst syndicate decided that the *Times* was a liability and likely to remain so. The paper was shut down in November 1960.

An intensive study of its 360 editorial and commercial workers was conducted by the Upjohn Institute (Ferman, 1963). The research findings show the reactions and unemployment experiences of these white-collar workers to be very similar to those of the blue-collar workers at Packard. The initial denial and disbelief, the anger and resentment, the depressing search for new jobs in a tight labor market, and the coming to terms with less attractive and less stable employment are already familiar. But these men and women had little previous experience with unemployment. Unlike the blue-collar workers in the automobile industry, they had no history of seasonal layoffs and prolonged job searches. The unfamiliarity of unemployment perhaps added to its pain and almost certainly to their difficulties in coping with it.

Six months after the shutdown, 50 percent of the *Times* employees were still unemployed and more than 40 percent had been so for the entire six-month period. The pattern of reemployment was predictable: The young were more successful than the middle-aged or old, the professionals somewhat more successful than those less skilled, the men more successful than the women, the college-educated more successful than their less educated co-workers. The jobs to which people moved were less stable and for some workers less rewarding; 17 percent of those who found new jobs lost them within six months, and 37 percent earned less than they had at the *Times*. Most who found jobs, however, continued to work at their previous skill level, and 19 percent earned higher wages. Meanwhile, for those without work the savings vanished, the bills piled up, and unemployment compensation provided for the daily necessities of life.

The crisis in the automotive industry 20 years later, in 1980, brought such problems again to many thousands of families, with costs that exceed conventional accounting. The research evidence is not yet available, but Michigan social indicators for 1979 and 1980 foreshadow what research is likely to show: Cases of child abuse increased 37 percent, substance abuse 10 percent, and suicide 27 percent.

SUMMARY

Although work is often a source of stress, it is also a source of gratification and most people know it. They recognize the demanding qualities of work but they believe that they would be worse off without it; in that sense, at least, the work ethic persists. Attachment to work takes many forms, however, depending on the nature of the job and of the person.

The strength of these attachments and the importance of work in people's lives are most apparent when work is denied. Studies of plant closings have documented the ramifying effects of job loss: the destrust and pessimism, the psychological depression, the sharply reduced living standard, the occasional suicide. They document also the recurrence of the unemployment experience, as men and women who lose their jobs in one place become the workers with the lowest seniority in another.

CHAPTER SIX

Job Loss and Health

The studies of the *Times* and Packard workers tell us a great deal about the experience of unemployment but less than we need to know about its effects on physical and mental health. The same must be said of the 17 studies of plant shutdowns available prior to 1977. In that year Cobb and Kasl published their research on the impact of the closing of two industrial plants in Michigan. Their study is unique in its concentration on the health effects of unemployment and in its inclusion of physiological measures along with personal interviews. It is unusual also in its longitudinal design—five sets of measures at regular intervals over a two-year period—in its initiation of measurement prior to termination, and in its inclusion of control groups of workers employed in four companies that operated in comparable settings and were in no danger of shutting down. In addition, there was a plant in which a shutdown had been anticipated but had not occurred during the two-year period of the study. Employment had shrunk in this plant, however, and workers employed there had been subjected to a number of internal transfers and job changes. They thus constituted a group that experienced the threat of job loss and the disruption of job change, but not the experience of actual unemployment.

WORK AND NUMBERS

The research of Cobb and Kasl on plant closings is available in two very different forms. Two books carry the title *Termination*, but their subtitles are different and they are very different books. Cobb and Kasl's own book, *Termination: The Consequences of Job Loss* (1977), is a technical report on the psychological and physiological condition of 100 men who lost their jobs when the Baker and Dawson plants closed in the 1960s. The effects of job loss are measured in centimeters, grams, and seconds wherever such measurements are possible; where they are not, psychological scales and indexes are used. Five sets of measurements were taken for each worker over a two-year period— at the early stage of anticipation when the plant closings were still rumors half-believed or denied, at the point of termination, and at intervals of 6 months, 12 months, and 24 months after the plants closed. Comparisons and statistical tests were made of the changes in well-being over this period, of the differences between the men who lost their jobs and those in similar plants who did not, and of the differences between those men who found other jobs quickly and those whose employment difficulties persisted. The findings fill a large technical monograph.

The other *Termination*—subtitled *The Closing at Baker Plant*—is a very different book. Its author, Alfred Slote (1969), is a professional writer and television producer who wanted to tell the story of the plant closings in a personal manner. With the encouragement of the researchers, Slote interviewed many of the workers, managers, and union representatives who had been at the Baker and Dawson plants when they closed. Thirty-one of those interviews, insightfully done and carefully transcribed, give the men who lost their jobs a chance to tell the story in their own words. Cobb, in the foreword, pays the book a researcher's tribute: "I consider this book one of the scientific reports of our study."

I share his opinion. The human narrative cannot substitute for the quantitative measurement and analysis of the technical report, but neither can the numbers substitute for the direct description of individual experience. In the following sections I summarize both sets of findings—first the numbers and then the words.

TERMINATION: THE CONSEQUENCES OF JOB LOSS

The Baker plant manufactured paint, primarily for the automobile industry. It was located in the large metropolitan area in which it had begun many years before as a small, family-owned business. In more recent years it had become part of large conglomerate, with corporate headquarters hundreds of miles away.

The Dawson plant manufactured lighting fixtures. Like Baker, it was a family-owned business that had been bought by a conglomerate corporation. Unlike Baker, the Dawson plant was located in a town of fewer than 3000 people. The plant was surrounded by farmland and many of the workers were part-time farmers.

One hundred workers were selected from these two plants, 46 from Baker and 54 from Dawson. For purposes of comparison, workers were selected from four other plants that were not threatened with shutdown—44 workers from urban plants and 30 from plants in rural settings. Finally, there were 28 workers from the Cryland plant, an urban manufacturer of automobile parts. Cryland was expected to close, but the closing did not occur during the two-year period of the research. As a result, the 28 high-seniority workers at Cryland form a group that was subjected to frequent job change and the threat of unemployment but not to job loss; they thus provide another basis of comparison for the workers at Baker and Dawson.

The *Termination* study is original in its systematic collection of health information during the closing of industrial plants.

There had been earlier evidence, however, that linked unemployment and illness (Cobb, 1969). A decade before, when the Studebaker plant in South Bend closed, the company agreed to continue health insurance coverage for the workers during their period of unemployment between jobs. The amount of the insurance premiums paid by the company was to be adjusted each month, according to the number of unemployed workers and the amount of service they required. Premiums rose so rapidly that the company questioned the arrangement, but the cause turned out to be increased incidence of illness and payment was continued. Staff members of the Detroit Metropolitan Hospital would have been less surprised; their admission rates go up regularly during times of layoff and model change in the automobile industry.

The workers who lost their jobs when Baker and Dawson closed were more fortunate than those who were caught in the closings of the *Detroit Times* and the Studebaker and Packard plants. The Baker and Dawson shutdowns were not accidents of the business cycle, and the displaced workers did not face a recession. Most of them eventually found new jobs that were as good as those they had held, and their average period of unemployment during the two years after plant closings was 15 weeks—considerably shorter than the period reported in the earlier studies. It seems likely, therefore, that the impact of the Baker and Dawson closings on the physical and mental health of the workers was correspondingly less severe.

The workers themselves rate the unemployment experience serious but not disastrous. On a scale that ranged from "hardly bothered me at all" to "changed my whole life," they put the impact of unemployment toward the heavy end—more than "somewhat disturbing" but less than "very disturbing." In comparison to other life events, good and bad, they thought that the plant closing demanded a lot of adjustment, less than divorce or death but more than most other life changes. It was about as important on the change scale as getting married, they said, but an unwanted change instead of one sought.

The Baker and Dawson workers did not blame themselves for their situation; there was little guilt. A few mentioned times of goldbricking and low productivity, but they placed most of the responsibility elsewhere—on management, the business situation, and the government, in descending order. In spite of the relative brevity of the unemployment experience and the lack of self-blame, the plant closings took a great toll in economic terms, in psychological effects, in physiological changes, and in health.

Economic Effects

At the time their plants closed, Baker and Dawson workers earned an average of $3.00 per hour, a rate that was then considered acceptable for industrial work. During the two-year period of the study they lost something over $5000 in wages as a result of unemployment, instability in their new jobs, and lower pay rates. Almost a year's wages, on the average, were lost.

The economic effects, and the workers' sense of them, were cumulative. The interviews taken during the phase of anticipated unemployment show no apparent sense of economic deprivation, but the situation changes sharply at the time of termination. The men reported difficulty in living on their reduced incomes, increases in debts and loans, and decreases or exhaustion of savings.

A sense of economic deprivation is a realistic response to the situation of unemployment, not a general apprehension about job change. Workers who went directly to new jobs without any period of unemployment did not report such problems, nor did the workers at the Cryland plant who were displacing those with less seniority.

Psychological Effects

The line between economic and psychological effects is not sharp. Concern about one's future security, apprehension about not getting ahead, and feeling that respect from others is somehow not forthcoming are all psychological responses that reflect economic realities. All these responses are related to the duration of unemployment, and none of them occurred among men who moved to new jobs without any intervening period of unemployment. Indeed, such men reported increased feelings of respect from others, as if they had proved themselves under difficult circumstances.

Mental Health and Well-Being

No one is going to care much about what happens, when you get right down to it. In spite of what some people say, the lot of the average man is getting worse, not better. You sometimes can't help wondering whether life is worthwhile anymore. Most people don't really care what happens to the next fellow. These days a person doesn't really know whom he can depend on. It is hardly fair to bring a child into the world the way things look now.

Those six statements make up an index of anomie, a state of uncertainty and pessimism about oneself and society, in which striving seems meaningless and the criteria of success uncertain. The men who lost their jobs at Baker and Dawson were more likely to consider those statements true than were the men in the plants that did not close. Furthermore, the anomie showed what the researchers called an unemployment effect and a job change effect; that is, the longer the worker was unemployed and the more jobs he tried before finding something stable, the more likely he was to agree with those six statements.

Cobb and Kasl used eight other indexes to measure mental health and well-being: depression, low self-esteem, anxiety and tension, physical symptoms, insomnia, anger and irritation, resentment, and suspicion. Again there were unemployment effects and job change effects. The men who were unemployed longer were more likely to be depressed, to have low self-esteem, to be anxious and tense, to be irritable, easily angered, and suspicious. Those who frequently changed job in the search for stable employment did not suffer the same losses in self-esteem or the same anxieties, but they did have similar feelings of anger, resentment, and suspicion, and they were more likely to report insomnia and psychophysiological symptoms.

Workers were presented with a series of cards, each containing a single statement, and were asked to indicate to what extent the statements applied to them.The more often a person had changed jobs after the plant closing, the more likely he was to say that the following kinds of statements were true for himself:

I am bothered by my heart beating hard.
I am bothered by dizzy spells.
I am bothered by shortness of breath when *not* exercising or working hard.
I often feel cold.
I do not feel healthy enough to carry out the things that I would like to do.
I often have a pain in my neck or back at the end of the day.
I have trouble staying asleep.
I have trouble falling asleep.

Physiological Changes

In a sense, physiological changes are the "hard" data of the research. Measures of cardiovascular and endocrine functions, of serum glucose and pepsinogen, and of the waste products of the kidneys are scientific in the conventional meaning of that word. It could be argued, however, that the subjective personal experience of job loss is the ultimate datum, that an elevation in blood pressure merely echoes what a worker has told us of tension and unexpressed resentment. More important than the priority of physiological, psychological, and economic measures, I believe, is their complementarity. All of them are necessary to describe fully the consequences of job loss.

Even during the period of anticipated unemployment, the Baker and Dawson workers showed higher blood pressure readings than the control groups in other plants. The elevated readings persisted during the period of unemployment and probationary reemployment and decreased as stable new jobs were acquired. The blood pressure readings of the workers in plants with stable employment were stable throughout the two-year period. The elevation in blood pressure was more persistent for workers whose unemployment was more extended or who reported more enduring stress. Moreover, it seems likely that the research findings on blood pressure understate the impact of job loss. Whenever a case of severe hypertension was suspected by the researchers, they referred the worker to a physician, on the grounds that individual well-being had priority over purity of research design. Twenty-nine such physician visits were recorded during the research period.

Cholesterol levels and pulse rates showed similar patterns—up from the time of anticipation to the time of actual job loss, and down from the time of unemployment to stable reemployment. Six months after the plant closings, the average pulse rate for the workers still unemployed was 83.1 beats per minute; for those who had found new jobs, it was 76.7.

Measures of neuroendocrine function have been used in the laboratory to indicate responses to stress, and they are used in medical diagnosis. Cobb and Kasl found that the effects of extraneous factors (coffee drinking, for example) made the analysis of unemployment effects extremely complicated. Nevertheless, some findings are unambiguous. The output of norepinephrine (noradrenaline), a measure of the activity of the adrenal glands, was higher for the Baker and Dawson workers than for the steadily employed workers in the control group. The excretion rates of norepinephrine for the Baker and Dawson workers had dropped significantly, however, when readings were taken 24 months after termination.

Protein-bound iodine is a measure of thyroid function. The analytic procedure is complicated and it was performed only during the earlier phases of the study (anticipation and termination). At those times, elevated readings (above 7.0 mg/dl) were recorded for 29 percent of the Baker and Dawson workers but for only 5 percent of the steadily employed workers in the comparison plants.

Functions Related to Other Diseases

Diabetes, peptic ulcer, and gout have long been considered diseases that have some relationship to stress. Moreover, for each of them there is a relevant indicator in the blood—serum glucose, pepsinogen, and serum uric acid. These measures showed distinctive patterns, some peaking earlier than others, but all of them reflecting some disadvantage or risk associated with the experience of unemployment. Among Baker and Dawson workers with few job changes and short periods of unemployment, only 8 percent had elevated blood sugar levels (130 mg/dl or more); among those with many job changes and longer periods of unemployment, 53 percent had one or more determinations of elevated blood sugar levels. Pepsinogen, the protein-digesting enzyme thought to be involved in the formation of peptic ulcers, showed a different pattern. Early readings were not elevated, but during the later phases of the transition from Baker and Dawson, workers who lost their jobs had higher amounts of serum pepsinogen than those who remained steadily employed. Serum uric acid, the immediate cause of gout, was significantly elevated among the Baker and Dawson workers during the stage of anticipated unemployment and dropped quickly as they settled into new jobs.

Diseases

Most diseases, fortunately, are rare events. Their rarity, however, makes it difficult to draw strong conclusions from a sample of 100 displaced workers. Of the 46 men who lost their jobs at the

Baker plant, two committed suicide, one attempted to do so, and one made suicidal threats. Four of the Baker men died of heart attacks, and three of the four had no previously known coronary disease.

During the first year after the plants closed, six of the men developed ulcers; among the steadily employed men there were two new ulcer cases during the same period, as measured by an index of symptoms (Dunn and Cobb, 1962). "Days of ulcer activity" were counted whenever people had stomach pain that woke them at night, came on before eating or several hours after eating, and then was relieved by taking food or milk. Men who lost their jobs had 17 times as many such days as those whose employment was stable. Peptic ulcer is rare among women, but three of the workers' wives were hospitalized for peptic ulcer within two months of their husbands' termination.

Arthritis, another stress-responsive disease, was measured by observation. The nurses who conducted the interviews examined the workers' hands and recorded the number of swollen joints. At the stages of anticipation and termination, 20 percent of the Baker and Dawson workers were found to have two or more swollen joints. These symptoms abated quickly for men who were unemployed for less than one month and moved to steady jobs. For those who did not, there seems to have been a later onset of arthritic symptoms. By the end of the research period, however, the former Baker and Dawson workers had no more arthritic symptoms than the steadily employed men in the comparison factories.

Hypertension appears early, under the threat of job loss, but in most cases recedes when stable employment is resumed. Twelve of the Baker and Dawson workers showed diastolic blood-pressure readings of more than 95 during the anticipation phase, but 10 of them returned to normal within the 24-month period, and the other two remained only slightly above normal. If one computes rates in spite of the small numbers, the rate of incidence of new cases of hypertension in continuous therapy among the Baker and Dawson terminees was 16 percent per year; no such cases occurred among workers in the comparison plants.

Other diseases occur still less frequently, and one can only note individual cases. The pattern of difference persists, however, between the men at Baker and Dawson and the men in the plants that did not close. Two new cases of diabetes were discovered, both among the Baker and Dawson workers. Eight of the Baker and Dawson men were reported to have been heavy drinkers; seven of them increased their consumption of alcohol during the time of unemployment and job change. Three cases of *alopecia areata* were reported, that strange stress-related disease in which hair falls out suddenly and in large patches. All three cases occurred among the men who had lost their jobs.

Self-reported information about health, disability, and the use of drugs (aspirin, tranquilizers, laxatives, etc.) shows the effects of stress even among workers who did not develop any of the diseases already discussed. The workers' individual health diaries record the days on which they "did not feel as well as usual." For 78 percent of the Baker-Dawson workers, the number of such days reduced as they regained stable employment.

"Days of disability" is another diary measure, which includes all days on which a worker did not carry on his usual activities for reasons of illness or injury. The number of such days—at home, in bed, or in the hospital—went down for the men who found stable new jobs and up for those who did not. Drug use was elevated at the stage of anticipated unemployment, but reduced by the end of the 24-month period of the research.

TERMINATION: THE CLOSING AT BAKER PLANT

Those are the numbers. Behind them are the intensely personal experiences of individual workers and their families. The quality of those experiences comes through in the 31 intensive interviews that Alfred Slote (1969) took with the men—managers, union representative, technicians, and hourly production workers—who had worked at the Baker plant.

Tom Morgan, who had been Baker's last manager of industrial relations and had begun an early retirement when Baker closed, showed Slote through the deserted buildings and explained the situation as he saw it:

> We could never stop production long enough to modernize. There were always crises, emergencies, production schedules to fill. And then one day it all caught up with us and they decided it was cheaper to start new somewhere else, with new people, than to rebuild the old. And that's where the hard part started. There wasn't a man who worked here who would have wanted to hurt this place. Their fathers worked here, their cousins, brothers, uncles, aunts. And when it was closing they just couldn't believe it. They wouldn't believe it. Hell, I should have known, and part of me didn't believe it either. Lloyd Shearer, who was the production manager at the end, didn't believe it either. It was just one of those things you didn't want to accept, I guess. Hell, it was like your family closing up on you. Your father throwing you out of the house. That's what this place was for forty years—one big family. . . .
>
> If you want to know what it was all about, go down to Arkansas and see Frank Robertson. He was the key to it all. I'm sure he'll talk to you. This plant was his family. He never had any kids of his own. And I'm sorry for him right now. (pp. 6 and 7)

Frank Robertson was plant manager during the closing of the Baker plant. He had worked in the plant for almost 40 years, moving gradually from laborer to technician, foreman, supervisor, and manager of industrial relations. As had Morgan, who replaced him in that job, Robertson retired when Baker closed. It was an early retirement. At the time of the interview, in the home he and his wife had rented in Booneville, Arkansas, he was 63 years old.

Because he was a manager, Robertson had not been included in the study of Baker and Dawson workers. He had, however, experienced a dramatic deterioration in his own health in the course of the plant closing, during which his weight dropped from 164 to 116 pounds. Slote gives the following account of Robertson's experience and their conversation together:

> Although he himself hadn't been sick one day in Detroit from the time he first arrived in 1925 to the final week of the plant closing in December of 1965, during that last week Robertson had attacks of pain and fever. His doctor examined him and, finding nothing, concluded it might be a virus and gave him some antibiotic. The fever did not clear up. Massive doses of antibiotic were administered. After a few days the fever went down but the stomach pain and bowel disorders lingered. Robertson went into the hospital for some tests the day after he retired in February 1966, and his illness was then diagnosed as mucous colitis—a persistently irritable colon which, according to medical literature, is mainly caused by tension associated with feelings of guilt. Robertson was told that his mucous colitis was caused by a reaction to the antibiotic. He was also told to expect to take medication for it for the rest of his life.

> Whether his illness is a reaction to drugs or to tension and guilt symptoms, Robertson concedes that today he feels guilty about his role in the closing of Baker. "I had a dream," he says. "I wanted to see that place grow. More than I ever wanted anything else. There were good people in that plant. People who were proud of what they were doing. No man, no company, ever had the right to expect the loyalty and sincerity that those men gave me. Right up to the last day. I hope to God they're all right now.

> "I guess some of them are not doing so well," he says softly, "but I'm not doing so well either. I was sick emotionally and physically when I came down here. I'd done just what I'd advised people for years not to do when they retired. I burned my bridges." He pauses again, to gauge the impact of these next words. "But there are just some bridges that won't burn, aren't there?"

> He stands up, walks over to the window overlooking the fields and the Arkansas hills, the sun making a pretty sunset behind them.

> "This acre of ground outside. It was a patch of weeds and a godsend. I worked on it from dawn to dusk when I first came down here. I worked on it trying to shut the long and sordid story of the

closing out of my mind. But at night it would catch up with me. I'd lie there in bed and go over it again and again, what had happened, why I'd done what I'd done, would I do it again? Could I have done it differently?

"Some nights would be worse than others. The nights were always worse than the days."

He turns and faces me, his face a study of resentment and relief.

"And now you've come down here, and you want me to go through it all again. I don't know if I can. I don't think you know what you're asking me to do. You say I can help other people. Yes . . . maybe I can. This thing is going to be a problem all over America. Plants closing, automation, progress, human beings. Maybe I can help." He pauses, and adds softly: "Maybe I can help myself too. Maybe the only way I'll get rid of this thing is to go through it again. All right, where do you want to begin? In Iowa where I was born? Or in Detroit where I went to work as a man? Or do you want to start when it really started, on Thurday, March 28, 1963, at 9 A.M. when I got a call from Harley Merritt, who was plant manager then." (pp. 9–11)

One of the men on Frank Robertson's mind was Ned Rockwell, who was chairman of the union bargaining committee when the Baker plant closed. When Alfred Slote interviewed him, Ned Rockwell was 41 years old and working as a laborer in a machine tool company. As Slote tells it:

Most of the time he operates a crane lifting heavy equipment around the plant. It is a dangerous and confining job and Rockwell hates it. He misses the freedom he had at Baker, he misses the importance of his role as Chairman of the union bargaining committee, he misses those days, especially during the closing, when he could roam the plant at will, tending to his men's wants and needs. At the machine tool shop where he now works, he is at the bottom of the barrel; he has only 18 months seniority.

"The funny thing is," he says, sipping a cup of tea, "I could have had a good job before Baker closed, but I stayed till the very end. I wanted to take care of my men. I wanted to ease the pain for them."

Rockwell is silent. Mrs. Rockwell, at the sink, watches him closely. "Do I miss Baker? Sure I miss it. I loved it there. A house'll stand up if there's love in it. A company too. I cursed it and cussed it. But I loved it. I loved the company. Nineteen years I worked there. Honest to God, I never did anything to hurt the Baker company."

He chokes off the sob that comes into his voice. (pp. 66–68)

Rockwell described his life since Baker closed as a series of defeats. The interview notes on his health are as follows:

Suffers from arthritis, ulcers, hypoglycemia, low blood pressure; has occasional dizziness and blackouts; absent from work month of January—"just couldn't get enough strength to go"; physician has recommended psychiatric consultation; Rockwell refuses to consider it; during the previous year, laid up six weeks with pneumonia. (p. 65)

Duane Paddleford, about the same age as Rockwell, has been more fortunate, somewhat to his own surprise. He had worked in the laboratory at Baker, spraying paint for chemical analyses, and now he is an inspector at a Ford assembly plant. He has begun to forget the time between, especially that cold March morning when he woke to find that much of his hair had fallen out during the night:

He felt the top of his head. "What the devil is going on?" He got out of bed and went to the bathroom mirror. He stared. His hair was falling—had fallen—out in clumps. Patches of bare skin shone in the overhead light. "My God, Norma, I'm sick."

He felt faint, a sudden rushing in his stomach. "I'm sick, Norma. I can't go to work."

But he'd never been sick in his life. Not a single day. He'd never been inside a hospital except to visit.

He lay in bed and stared at the ceiling and every once in a while he'd pass his hand over his scalp and feel the skin and once another clump of hair came off in his hand and he was scared and frightened, because he knew he was sick now, real sick. He was sure he was going to die.

Duane saw his doctor immediately.

"He gave me a thorough exam . . . and said he guessed this condition of mine might be caused by some emotional problems, and I said I didn't have any emotional problems that I knew of and Norma [Mrs. Paddleford] who was there said: 'Baker's closing.'

"Ralph [the doctor] says, 'Oh.'

"And I say: 'Naw, they're just bluffing about that. That's what I think.'

"And Ralph shakes his head and says: 'You may think you think that, but deep down you're probably thinking something else.'

"He gave me some pills to take. Tranquilizers, I guess, and told me to stay home a couple days and take it easy and that I could go back to work when I felt like it."

"I kept doing everything I'd been doing, and I guess the pills helped. Because about a month later Norma looks at me across the table and she says: 'Duane, your hair's starting to grow back.'

"And Susie she kisses me. I almost cried then for real. This was in March of '64. By December all my hair was back and I was almost ready to go to the barber's again. And by September of '65, I'd pretty much forgot the whole thing. I was off the pills too. . . .

"And then do you know what happened? The next month? October of 1965—it started all over again. I woke up one morning and my hair was falling out again. I couldn't believe it. Not again. It was awful, even worse the second time. I felt like crying. You didn't think it could happen again. Not twice. But it did. Of course, in one way I knew it wasn't so bad. I knew I wasn't going to die; I knew it was my nerves and I knew what was making me so nervous—Baker. Just the week before I had got the letter telling me I was to be terminated the following Friday. And I guess that did it. Started it all up again. I guess I was more worried about finding a new job than I knew.

"I went back to Ralph and he put me on the pills again. Three days after I was laid off at Baker I signed on at Ford, and I haven't

missed a day's work since. I've even been promoted to inspector—
and look at my head! It started growing back two weeks after I
started in at Ford, and now I go to the barber's twice a month.
He's got his work cut out for him too. Isn't it something how you
can be worried about a thing and never know it till your hair falls
out?" (pp. 124–126)

SUMMARY

Most studies of unemployment are cross-sectional and retro-
spective; that is, they are based on interviews taken at some one
point in time after the job loss has occurred. The research of
Cobb and Kasl (1977) on plant closings in Michigan is unusual in
several respects: Workers who did not lose their jobs were included in the study, as well as those who did; five interviews were
conducted with each worker over a two-year period; and physiological measures were obtained along with self-reported information. The research results show that the costs of unemployment are physical as well as economic and psychological. They
show also that the threat of unemployment triggers some physiological changes long before actual job loss occurs and that most
physical indicators return to normal, sometimes slowly, after
new job stability is attained.

CHAPTER SEVEN

Workers and Jobs: Goodness of Fit

My argument thus far can be quickly summarized. First, most people want to work, and not only for economic reasons; the deprivation of work is a stressful and damaging state (Chapter 5 and 6). Second, all work is not of one kind, and how workers react to their jobs depends on the nature of those jobs (Chapters 2, 3, and 4). If policy were to be deduced from the evidence we have reviewed, two principles would dominate; the entitlement to work, for those willing and able, and the entitlement of workers to a "good job." Job entitlement and job improvement might be the shorthand terms.

The second of those phrases is more ambiguous than the first, because all the criteria of a good job do not converge and a job that is good for one person may not be good for another. Let us consider a good job for a person to be one that he or she can do well, can do without physical or mental damage, and can do with satisfaction. Doing a job well is usually expressed in terms of quantity and quality of product, and management attempts to assess performance in those terms. Whether a job is done with satisfaction and without physical or mental damage is less often measured, but measures have been made and evidence is accumulating, as we have seen.

All three criteria of a good job—one that a person can do well, can do without damage, and can do with satisfaction—must eventually be considered in the individual case. People differ, people change, and no one set of job satisfactions is ideal for everyone at all times. The characteristics of the job must therefore be assessed in relation to the characteristics of the man or woman who holds it. Assessment becomes a problem in goodness of fit between the worker and the job.

THE MINORITY VIEW

The recognition of individual differences by no means negates the general trends described earlier. People do tend to agree about what makes a job good, as defined by the eight dimensions that begin with the content of the task itself and end with the larger organizational context in which the task is performed (Chapter 3). General tendencies, however, imply that there are minorities as well as majorities, and the voice of the minority has occasionally been heeded in research on work.

In an early and widely quoted experiment (Morse and Reimer, 1956) on the distribution of power in organizations, four divisions of clerical workers in a large corporation were involved in two different and opposite experimental treatments. In the "autonomy program" the locus of authority was moved downward; successive levels of management, beginning with the vice-president, delegated areas of authority they had previously held. The clerical workers themselves made group decisions about rest periods, the handling of tardiness, methods of work, and the like. In the "hierarchically controlled" program, on the other hand, control from above was tightened; decisions formerly made by first-level supervisors were made by second-level managers, and decisions the latter had formerly made were now made by heads of departments. Employees were less involved than before in regulating and controlling their own activities.

Somewhat to the surprise of the experimenters, productivity

increased in both these experimental programs. The other reactions of employees to the programs, however, were much as predicted. Self-actualization increased in the autonomy program and decreased in the hierarchical program. Relationships with supervisors improved in the autonomy program and deteriorated in the other. Liking for the company followed the same pattern. Most clerks in the autonomy program felt they were gaining much from it, wanted it to last indefinitely, and did not like the other program. Most clerks in the hierarchical program felt that the company was gaining; they wanted it to end immediately and expressed a liking for the other program.

In both programs, however, there was a dissenting minority. A few clerks in the autonomy program felt uneasy with their increased freedom and responsibility and disliked the process of group decision. They expressed a preference for the preexperimental days when a supervisor had told them what to do, how to do it, and when it had to be done. And a minority in the hierarchical program said they were more satisfied now that firm standards were set from above and there was no room for excuses or discussion of alternatives.

Tannenbaum and Allport (1956) explained these different reactions in terms of personality differences. They used short written tests to measure the personality traits of the workers involved in the experiment, and on the basis of these tests judged whether the program to which the workers had been assigned was "suited" or "unsuited" to their personalities. Thus a person with strong needs for independence and group affiliation would be considered "suited" if she had happened to be in a division assigned to the autonomy program and "unsuited" if she had been in one of the divisions involved in the hierarchical program. People thus suited to their programs wanted them to continue longer and expressed increased satisfaction with them; these findings held for both programs. Fewer people were suited to the hierarchical program, but those that were liked it.

Vroom (1960) reported similar findings from a study of supervisory practices in a package delivery company. The company operated nationally through local units in major cities. Each such unit had a central depot or station where packages were brought and sorted according to destination and from which they were delivered by trucks covering designated territories. Performance and satisfaction levels were found to be higher in those units where the supervisory style was participative than in those where it was not. The strength of this relationship, however, depended on the personality of the individual employees. Vroom found that the feeling of participation had its greatest effects among those whose need for independence was strong and whose tendencies toward authoritarianism were weak. The effects of participation were small, but not negative, among people who had little need for independence and had strong authoritarian tendencies.

Generally, the evidence confirms what intuition might have suggested in the first place: Job satisfaction depends on the characteristics of both the job and the individual who holds it. The evidence also suggests that the situational factors are probably the more powerful. If we must predict job satisfaction from only one kind of information, we can do best by basing our predictions on the characteristics of the job itself. We can improve those predictions, however, by considering the characteristics of the individual at the same time.

THE P-E FIT MODEL[1]

Researchers agree that to understand how a person will react in a particular situation, facts about the situation and facts about the person must somehow be combined. They do not agree about how the combination of such diverse characteristics should be made. Various proposals have been advanced and various models constructed—subtractive, multiplicative, and miscellaneous. One of these, the P-E (personality–environment) fit model

---

[1]This description is adapted from French, Rodgers, and Cobb (1974).

(French, Rodgers, and Cobb, 1974), is particularly useful for our purposes. It is internally consistent, it makes specific predictions, and it deals with the overall adjustment of the individual to a given situation. Moreover, it has been the vehicle for several research investigations involving the relationship between job characteristics and worker well-being.

Basic Concepts

The basic idea of the P-E fit model is that individual adjustment consists of goodness of fit between the characteristics of a person and the properties of that person's environment. A person's environment might include, for example, the work situation, the family arrangement, and the neighborhood.

A further distinction is made between objective facts—things as they really are or might be assessed by an objective, expert observer—and subjective facts—things as they are perceived by the person whose adjustment or well-being we wish to understand. Thus for each relevant property of the person's environment there will be an objective measure and a subjective measure; for each individual there is in short an objective environment and a subjective environment, the world as it really is and the world as that person perceives it.

A similar distinction must be made for characteristics of the person; for each characteristic there is an objective measure and a subjective measure. Thus for each individual there is an objective person—the relevant characteristics of that individual as objectively determined—and a subjective person—the same characteristics as they appear to that individual.

These distinctions between objective and subjective properties imply two measures of adjustment. Objective adjustment (objective P-E fit) is the goodness of fit between objective properties of the environment and the corresponding objective characteristics of the person. Subjective adjustment (subjective P-E fit) is the goodness of fit between the subjective properties of the environment and the corresponding subjective characteristics of the person. These four basic concepts and the two measures of adjustment are represented in Figure 7.1.

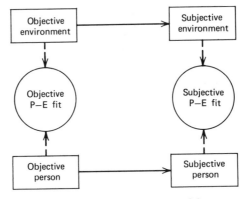

Figure 7.1.   P–E fit: core model.

Example and Elaboration

Let us take an example, a case of goodness of fit between a job requirement and the corresponding ability of the person holding the job. Suppose that the job consists of manuscript typing and calls for an average typing speed of 50 words of finished copy per minute. Suppose that the person holding the job can average only 30 words per minute. Objective P-E fit is poor in this respect; there is a discrepancy of -20 words per minute between the objective job requirement and the objective ability of the typist.

The typist may not see it that way, of course. He or she may have a defensively reduced view of the job requirements and a defensively exaggerated view of his or her typing speed. Let us assume that both these defenses have occurred, and that the typist reports that the job requires only 40 words per minute (instead of 50) and that he or she types 40 words per minute (instead of 30). Subjective P-E fit is now perfect; the typist thinks the job requires 40 words per minute and that he or she types 40 words per minute.

Such defenses may or may not be effective. They may, for example, relieve an immediate anxiety about inadequate performance at the cost of greater anxieties in the future. Whether the typist holds the job for long is likely to depend on objective P-E fit and not on subjective fit. But regardless of that outcome, the defenses involve a cost in mental health and well-being. In this example the typist understates the requirements of the job; this is a distortion of reality, and contact with reality is one criterion of mental health. The typist has also exaggerated his or her own ability. This is distortion of another kind; it reduces the accuracy of self-assessment (sometimes called accessibility of the self), which is another criterion of mental health.

We can imagine other ways of dealing with an initial discrepancy between an objective job requirement and the corresponding ability. The typist might have acted to change the requirement, perhaps by persuading the supervisor that 50 words per minute of error-free copy was an unreasonable demand week in and week out. Failing that, the typist might have acted to increase his or her typing speed, perhaps by enrolling in a special course for manuscript work. Such actions to improve objective goodness of fit, either by changing the relevant property of the situation or of the person, are called *coping*. We can further distinguish the two kinds of coping provided by this example. Increasing objective P-E fit by bringing about a change in the environment (in this case, persuading the supervisor to redefine the job requirement) we would call *environmental mastery*. Increasing objective P-E fit by changing one's own characteristics (in this case, increasing typing ability) we would call *adaptation*.

These concepts can be incorporated in the core model of P-E fit presented in Figure 7.1. The full model thus consists of 10 concepts and the specified relationships among them, as shown in Figure 7.2. From objective and subjective goodness of fit, as defined by this model, we would expect to predict the strains and satisfactions experienced by a person and the ramifying effects on health.

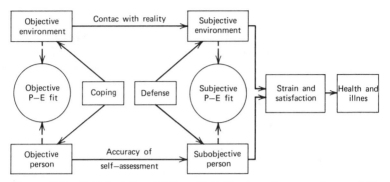

Figure 7.2.   P–E fit: full model [from V. R. Harrison, Person-environment fit and job stress, in C.L. Cooper and R. Payne (Eds.), *Stress at Work* (New York: Wiley, 1978)].

Supply and Demand

Goodness of fit between a person and a job can be thought of as a problem of supply and demand. In the example of the typist, the demand is the requirement of the job for a typing speed of 50 words per minute. The supply is the ability of the typist, also expressed in words per minute. For any job there will be a set of such requirements, and for any person a set of corresponding abilities. If we are to compare any demand and the corresponding ability, we must define and measure them in commensurate terms, that is, with the same unit or scale.

The example of typing speed was chosen for its familiarity and convenience, since both the job requirement and the relevant ability are customarily expressed as words per minute. Other characteristics, however, can be treated in commensurate terms. For example, we might rate the complexity of a task on a 7-point scale ranging from very simple to very complex. The ability of a person to perform complex tasks could be rated on the same scale.

In these cases the demand is in the job and the supply is in person. All goodness-of-fit problems, however, are not of that kind. It is not always the job that demands and the person who supplies. People have needs, goals, and preferences as well as abilities and skills. And jobs are sources of opportunities and rewards as well as demands and requirements. For example, a person may have a strong need for affiliation—being a member of a group of people who interact with each other and value one another's company. The job can be rated by the extent to which it provides opportunity for such interaction. In this case the job becomes the source of supply, meeting or failing to meet the corresponding need of the person who holds it.

Too Little and Too Much

When a person has too little skill or ability to meet the requirements of a job or when the opportunities or rewards of a job are too meager to meet the needs of the person, strain is the predictable consequence. In both cases demand exceeds supply, and the objective fit is less than perfect. The greater the lack of fit, the greater would be the predicted strain.

It is not equally clear what predictions should be made when supply exceeds demand. What if a person has more ability than the job requires, or what if the job offers more opportunity for interaction than the person needs? One can imagine cases in which too much of some supply would be as stressful as too little. For example, having more responsibility than one can handle is certainly stressful but having less than one wants is almost certain to be boring or frustrating. The predicted relationship between responsibility and strain, therefore, is U-shaped; strain is at its minimum when the amount of responsibility corresponds to the needs of the individual, and strain increases more or less symmetrically when responsibility is either too much or too little for the individual.

On the other hand, consider the potential stress involved in the opportunity for interaction with others. Too little such opportunity on the job is experienced as stressful, but what if the opportunity for interaction is increased so that it exceeds the needs of the individual? Is that stressful? We do not know, but it seems unlikely. Informal interaction at work is not compelled, and if there are more opportunities for it than a worker wants, he or she presumably can ignore some of them. The curve illustrating the relationship of interaction opportunities to strain would assume the familiar monotonic shape; that is, the greatest strain would be reported when no such opportunities existed. Strain would decrease as opportunities increased and reach its lowest point when opportunities equaled needs. As opportunities increased beyond needs, however, strain would remain at the same low level. Increasing opportunities for interaction, once one has all one wants, would be neutral in effect. The marginal utility, as the economists might call it, of interaction opportunities might drop to zero but it would not become negative.

In short, we expect that the relationship of goodness of fit to strain will be U-shaped for some characteristics and monotonic for others, as shown in the two hypothetical curves in Figure 7.3. Which curve will hold for which characteristics is a question requiring more research. In both curves perfect goodness of fit is at the zero point; deficiencies are indicated by negative values on the scale of fit, and excesses are indicated by positive values.

Few of the major characteristics of jobs (Chapter 3) have been studied in ways that allow us to judge goodness of fit between the job and the person. Researchers seldom design scales that measure supplies and demands in comparable terms, and thus the

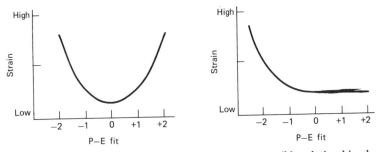

Figure 7.3. Hypothetical curves illustrating two possible relationships between P-E fit and strain.

judgment of too little and too much can seldom be made. A few studies have been done, however, that provide some evidence of the importance of goodness of fit between the person and the job.

JOB COMPLEXITY AND STRAIN

In one such study (Caplan et al., 1975) more than 2000 workers in 23 occupations were asked about their needs, their preferences, and the content of their jobs. Measures of psychological strain (dissatisfaction, depression, anxiety) were obtained from all these workers, and measures of physiological strain (blood pressure, heart rate, cholesterol, and others) were obtained from a subsample of them.

The main measure of task content was an index of job complexity, which included questions about whether the job involved an imposed work routine, whether it required the performance of only one task or of several different tasks, and whether in doing the job the worker dealt with a number of other people. One of the measures of strain was the Zung scale (Zung, 1965), an index of depression that had been validated by means of clinical assessment. The relationship between job complexity and depression is shown in Figure 7.4. It is a weak relationship, but the direction is clear; feelings of depression are most common among workers doing jobs of low complexity and least common among workers doing highly complex jobs.

In this study workers had also been asked about their preferences for simple or complex tasks, so it was possible to develop a goodness-of-fit score for each worker by subtracting the amount of complexity preferred by the individual from the amount provided by the job. The relationship between this score (P-E fit of job complexity) and the depression scale is shown in Figure 7.5. It is a stronger relationship than that shown in Figure 7.4, and it has the expected U-shaped configuration. It shows that too much job complexity is at least as stressful (depressing) as too

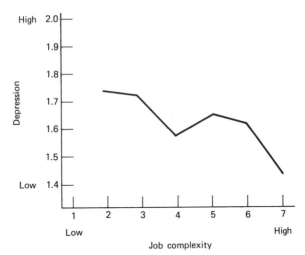

Figure 7.4. Relationships between scores on depression and scores on job complexity: eta = .14 (N.S.), N = 318 men from 23 occupations [from V. R. Harrison, Person-environment fit and job stress, in C.L. Cooper and R. Payne (Eds.), *Stress at Work* (New York: Wiley, 1978)].

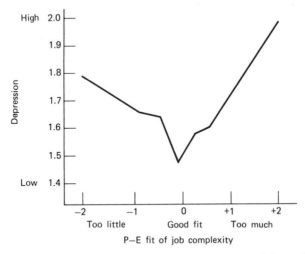

Figure 7.5. Relationship between job complexity P-E fit and depression: eta = .26, p <.002, N = 318 men from 23 occupations [from R. D. Caplan, S. Cobb, et al., *Job Demands and Worker Health: Towards a Full Share in Abundance* (Washington, D.C.: U.S. Government Printing Office, 1975), p. 91].

112

little and that feelings of depression are better understood in terms of goodness of fit than in terms of job complexity alone.

WORK LOAD AND STRAIN

The index of work load measures the quantitative demands of jobs and thus complements the index of job complexity, which measures their qualitative demands. Work load does not show the same U-shaped relationship to measures of strain, however. The explanation for this involves occupational differences; too much work (overload) is stressful in any occupation, but whether too little work is stressful depends on the occupation itself.

These differences are illustrated in Figure 7.6, which shows

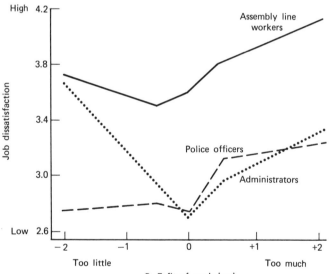

Figure 7.6. Relationships between job dissatisfaction and P-E fit on work load: for assembly line workers, eta = .31, $p < .04$, $N = 1.02$; for administrators, eta = .38, $p < .001$, $N = 239$; for police officers, eta = .29, $p = .05$, $N = 108$ [from V. R. Harrison, Person-environment fit and job stress, in C. L. Cooper and R. Payne (Eds.), *Stress at Work* (New York: Wiley, 1978)].

the relationships between job dissatisfaction and the P-E fit of work load, that is, goodness of fit between the ideal amount of work as reported by the worker and the actual amount of work required by the job. That relationship is shown separately for three occupations—assembly line workers, police officers, and administrators.

Several differences among these occupations can be seen. First, the height of the curves is different; job dissatisfaction is highest among assembly line workers. Second, the configuration of the curves is different. Administrators show the familiar U-shaped curve, not quite symmetrical. They are dissatisfied with too little work as well as with too much; in fact, it appears that they may be more dissatisfied with too little work than with too much. Administrators with too much to do may feel important; administrators with too little to do may feel superfluous and perhaps worry that others may judge them to be so. Assembly line workers show only slight increases in dissatisfaction when they have too little to do, and policemen show no increase whatsoever when there is too little to do. And for all three occupations the zero point, at which the amount of work to be done fits the needs and abilities of the worker, approximates the point at which dissatisfaction is minimized.

We can only speculate about the aspects of these occupations that account for such differences. For example, we know that administrators rank high in intrinsic job satisfaction; they say that the content of the work itself is challenging and rewarding. If there is too little work, then perhaps there is also too little challenge and reward in its accomplishment; hence the increase in job dissatisfaction. Assembly line workers report little intrinsic interest in their work and presumably would feel no such pangs when there is too little work to do. They might, of course, feel bored with their idleness or worried about being laid off if the dearth of work were to continue. Police officers may be a special case in that their work is a combination of routine underload and crisis overload. Police work is also an occupation in which underload is in a sense reassuring; all quiet means all is well.

OTHER GOODNESS-OF-FIT MEASURES

In two studies of managers, scientists, and engineers (French, 1973) goodness of fit was measured on several dimensions, including work load, responsibility for others, role ambiguity, relations with subordinates, and participation in organizational decisions. Goodness of fit on each dimension was then examined in relation to various measures of strain, including job dissatisfaction, feelings of job-related threat, anxiety, depression, and somatic symptoms.

The general proposition about goodness of fit is sustained in these studies: Strain is minimized and satisfaction maximized when the demands and opportunities of the job fit the abilities and needs of the individual. For 9 of the 11 job characteristics measured in the first of these studies, goodness of fit was significantly related to satisfaction. There were variations on this basic theme, however. For example, people are most satisfied when the degree of clarity or ambiguity of their job fits their needs, but the curve that illustrates this relationship is asymmetrical; most of these scientists, engineers, and administrators encountered more ambiguity than they would have wished. Goodness of fit with respect to quantitative work load, participation, and responsibility for others was also asymmetrical; most of the people in this study wanted more work, participation, and responsibility than they had. Symptoms of depression, somatic complaints, and feelings of threat were minimized at the zero point on all these dimensions—that is, when goodness of fit between the characteristic of the person and property of the job was perfect.

THE STATISTICAL TEST

The goodness-of-fit hypothesis, like other scientific propositions, should pass at least two tests. First, it should explain something, not merely be true by definition. Second, it should provide a better explanation than other and simpler propositions.

Here, the first test seems easier than the second. Admittedly, the goodness-of-fit hypothesis at its simplest becomes almost a tautology: Pepole are less strained and less dissatisfied when they get what they need and want. We have seen, however, that the story is more complicated than that. The amount of responsibility that maximizes self-esteem, for example, may be too much for comfort in other respects, and the amount that people say they want may be more than suits them when they get it.

The second test is best stated as a problem in statistical prediction; for each characteristic of interest, we must ask whether the appropriate measure of P-E fit predicts the various criteria of strain better than does the corresponding measure of the personal characteristic alone, better than the measure of the job characteristic alone, and better than both of them together.

Three such studies permit us to make these comparisons, and all of them offer some support for the goodness-of-fit hypothesis. House (1972), in an analysis of characteristics of job satisfaction, found that about one-third of the time (5 characteristics of 16) the goodness-of-fit measures added significantly to the explanation provided by measures of personal characteristics and job properties taken separately. Kulka (1976), in a study of high-school students, found that goodness-of-fit measures added to the explanation provided by personal and situational characteristics about one-fifth of the time (44 of 205 relationships studied).

The most complete comparison of this kind is provided by Harrison (1976). Working with a sample of 318 men in 23 occupations—actually a subsample of the larger study of 2010 men described earlier—Harrison examined the effects of goodness of fit along four dimensions: job complexity, work load, responsibility for others, and role ambiguity. For each of these dimensions, the effect of goodness of fit was studied in ralation to several measures of strain. In 18 of 27 such relationships strain was predicted better ty the P-E fit measures (that is, by goodness of fit) than by the job characteristic alone, the corresponding characteristic of the person alone, or the two combined.

The advantage of the goodness-of-fit measures is greatest when the relationship between a job characteristic and a measure of strain is U-shaped, that is, when too much of the characteristic and too little of it are equally stressful. Table 7.1 presents data for six such relationships.

Table 7.1   Correlations and Multiple Correlations between Strains and E, P, and Misfit Indexes of Job Complexity[a]

	Correlations[b]				Multiple Correlations[b]
	(1)	(2)	(3)	(4)	(5) E and P with Additional Effects of Misfit[c]
Strains	Environment (E)	Person (P)	Misfit (E-P)	E and P	
Job dissatisfaction	−.31	−.30	.47	.34	.50
Boredom	−.51	−.34	.51	.51	.61
Somatic complaints	NS	NS	.16	NS	.18
Anxiety	NS	NS	.21	NS	.25
Depression	NS	−.12	.22	NS	.23
Irritation	NS	NS	.15	NS	.20

[a]From Harrison, 1976. The data are based on a study of 318 men from 23 occupations. For all correlations presented, $p = .05$.

[b]NS 4 not significant.

[c]The addition of misfit increases the multiple correlation significantly.

The five numbered columns of Table 7.1 show various measures of complexity as a characteristic of jobs and as a need or preference of persons. The left column lists six measures of strain: job dissatisfaction, boredom, somatic complaints, anxiety, depression, and irritation (feelings of anger, aggravation, annoyance, etc.). All the entries in the table are correlations, indicating the power of the various measures of job complexity to predict the different measures of strain.

Column 1 is based on measures of the job: its complexity or simplicity is a characteristic of the worker's environment. Column 2 is based on parallel questions regarding the person—the preferences of workers for complex or simple jobs. Column 3 is a subtractive measure of goodness of fit (misfit)—the difference between the complexity of a person's job and that person's need (preference) for complexity. Column 4 shows the combined power of job complexity and worker preferences as predictors of strain, and column 5 adds to these two overall predictors the measure of misfit.

The table demonstrates that we can understand the reactions of workers to their jobs more fully if we use the goodness-of-fit concept and measure the characteristics of the job in relation to the needs of the worker. Some information is gained from the job characteristics alone; extremely simple and routine jobs (low in complexity) generate job dissatisfaction and boredom. Whether they also evoke other symptoms, somatic and emotional, depends on how large the gap is between the complexity of the job and the need of the individual.

GOODNESS OF FIT AND ORGANIZATIONAL CHANGE

The pragmatic implications of the P-E fit model are in some ways more difficult to respond to than the theoretical questions. Let us suppose that we wish to reduce the strains and dissatisfactions associated with work; what help can we get from research on goodness of fit?

The first answer to that question, I believe, comes from the asymmetry in some of the curves discussed earlier. Most people want more job complexity, more participation, and more clarity, not less. It follows that overall goodness of fit can be improved by increases in these job characteristics, even if each change is not tailored to the preferences of the individual. Moreover, the changes need not be randomly introduced; they can be concentrated in jobs that are most deficient to begin with.

A second implication of the goodness-of-fit approach is that strains are especially likely to be reduced by organizational changes that permit individuals to modify the characteristics of their own job—to engage in what social psychologists call role elaboration. Jobs well up in the organizational hierarchy are shaped in substantial ways by their occupants. That is why the appointment of a new company president is watched with such interest by rank-and-file employees as well as by vice-presidents. It is a challenge for organizational designers and changers to build some such flexibility into jobs at other levels, so that within limits a worker may set his or own pace, participate in organizational decisions or not, take on more responsibility or leave it alone. Goodness of fit then becomes a matter of fine tuning by the individual.

A third implication of the goodness-of-fit approach might be thought of as an extension of the familiar procedures of selection and placement. Two kinds of fit between the person and the job have already been discussed—the fit between the person's abilities and the job's requirements, and the fit between the person's needs and the job's opportunities. Hackman and Suttle (1977) call these two kinds of fit congruence and illustrate them as shown in Figure 7.7. Conventional selection and placement pro-

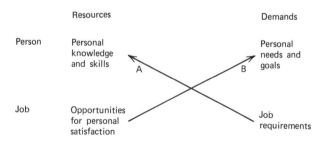

Figure 7.7. Congruence between person and job [from J.R. Hackman and J.L. Suttle *Improving Life at Work* (Santa Monica, Calif: Goodyear, 1977), p. 115].

cedures address themselves to only one of these dimensions (arrow *A*), because poor fit on this dimension means poor performance and reduced organizational effectiveness. The P-E fit model implies similar attention to the other dimension (arrow *B*), because poor fit here means increased strains and reduced well-being. Maximizing fit simultaneously along both dimensions becomes the goal of organizational change and design.

SUMMARY

Earlier chapters have been concerned mainly with general tendencies. This chapter is concerned with individual differences. Few research studies report relationships between experimental changes at work and personality characteristics of workers; those that do show significant individual differences. Not all individuals respond in the same way to the same conditions of work. Majority responses can be predicted from a knowledge of job characteristics, but to predict more powerfully requires concepts that include characteristics of individuals as well. Person–environment fit is such a concept, and researchers have begun to use it or its equivalent. Such research indicates that most people want more complex tasks and more participation in decisions, for example, but that some people do not. The implications for organizational change are briefly considered.

CHAPTER EIGHT

People Changing: Strategies of Adaptation

Millions of people do not have jobs that enhance their health, their satisfaction, or their well-being. Some are unemployed; many more have jobs that do not fit their needs. The problems of unemployment have already been discussed. What can be done to improve the goodness of fit between workers and jobs?

There are, I believe, three basic answers to that question, each representing a different orientation to policy and implying a different program of action:

1. *Living with it.* We can tolerate the discrepancy and accept the idea that work is boring for many people, damaging to many others, and that the consequent strains must be absorbed or compensated for in other parts of society.

2. *Changing people.* We can improve the goodness of fit between people and jobs by changing people in either of two senses—changing their assignments or changing their attributes. The former requires that we become more adept at selection and

121

placement, transfer and promotion, and perhaps at termination and "counseling out." The latter involves information and education, training, counseling, psychotherapy, behavior modification, and programmed group encounters of various kinds. In both strategies, people moving and people changing, the primary target of change is the individual.

3. *Changing jobs.* We can improve the goodness of fit between people and jobs by changing the jobs—that is, by changing the stiuations in which people work and the tasks they are expected to perform. The primary target of such change may be the job itself, its rewards, or the larger organizational structure within which jobs are defined.

Each strategy has its benefits and costs, some obvious and some hidden. Each involves certain assumptions and expresses certain value preferences. Finally, each presents distinctive problems of implementation, some of which have been studied extensively but most of which have not. In this chapter I attempt to assess two of these three approaches to the quality of working life.

LIVING WITH IT

The idea that the negative qualities of work are essentially unavoidable is as ancient as the biblical curse with which Adam and Eve were cast out of Eden: "Cursed is the ground for thy sake; in toil shalt though eat of it all the days of thy life" (Genesis 3:17). Since that time, the acceptance of drudgery has been argued on many other bases, but usually by people engaged in relatively pleasant kinds of work.

Those who counsel others to accept bad jobs uncomplainingly sometimes attempt to lighten the burden of their advice by arguing that nonwork activities are insulated from those of work, or even that people choose nonwork activities that contrast with and counteract the demands of the job. It is an appealing thought—the machine operator turned artisan after hours or the daytime taker of orders emerging as the evening leader of community organizations.

There is some basis for this notion. To the extent that jobs are well paid, hours of work are short, tasks are not too fatiguing, and the larger society is not strongly status-defined, one might expect that the constraints of the job would not check activities away from the job. Some of these conditions are met, at least to some degree, in the United States and in some other affluent industrial societies.

Nevertheless, the weight of the evidence contradicts this reassuring view. If we must choose among oversimplifications, the principle of consistency holds true more often than that of compensation or complementarity. The worker hooked to the machine by day is more likely to spend the evening hooked to the television set than engaged in some compensatory and creative leisure activity. Neither the principle of consistency nor that of complementarity, however, is sufficient to explain the complex relationships between work and nonwork. Some aspects of work—such as the number of hours spent at it—inevitably reduce the time available for nonwork activities; each of us has only 168 hours each week. For other aspects of work, however, the conclusions are more problematic. For example, is individual energy like time or is it a state of mind that, turned on at work, persists when work is through? Is leisure more satisfying when work is not?

Research on such questions has been fragmentary and specialized, but some recent work (Staines, 1977; Staines and Pagnucco, 1977) has brought together most of the fragments and added a study of 651 workers in a wide range of occupations and industries. This research uses a framework in which work and nonwork are compared in three main categories—type of activity, degree of involvement, and subjective reaction. As Table 8.1 illustrates, nine areas of work–nonwork relationships are thus generated and three of them are especially relevant for exploring the competing hypotheses of consistency, compartmentalization, and complementarity.

Table 8.1   Relationships between Work and Nonwork:
Activity, Involvement, and Reaction

| | Nonwork | | |
Work	Types of Activity	Degree of Involvement	Subjective Reaction
Types of activity	1	2	3
Degree of involvement	4	5	6
Subjective reaction	7	8	9

Types of Activity

Studies of activity patterns on and off the job favor the consistency hypothesis. Research on Canadian workers (Meissner, 1971) shows that those who are technically constrained at work, machine-paced or confined in space, are less likely to engage in leisure activities that require initiative, planning, or decision making. Similarly, workers on jobs that prevent their talking with others are less likely to participate in sociable activities off the job—visiting, talking, or going on family outings. In Swedish sawmills and lumber-trimming plants, workers on machine-paced jobs with short operating cycles and limited freedom of movement participated less in off-the-job organizations, including their labor unions (Gardell, 1976). Other research (Parker and Smith, 1976) shows the consistency pattern for leadership activities; people on jobs that required leadership or supervision of others were more likely to participate actively in community organizations and associations.

Whether these findings are interpreted as the effects of jobs or as the result of selective processes, they offer little support for notions of complementarity between type of work activity and type of nonwork activity. The sole exception is the suggestion that workers in physically exhausting jobs tend to avoid physically demanding leisure activities. Physical energy, like time, must have limits for any individual and any given period; what is spent in one place is not available for use in another.

Similar findings came from the Survey Research Center study of 651 workers in five establishments—a hospital, a printing company, two manufacturing companies, and a corporation engaged in research and development. Correlations between work and nonwork activities were generally low, suggesting some independence or compartmentalization of work from nonwork activities. But most of the correlations were positive, and all those large enough to be statistically significant were positive. To the extent that the type of activity in one life sector (nonwork) is influenced by the activity in another (work), the tendency is toward consistency. We tend to become what we do.

Degree of Involvement

Here again there is some evidence for consistency; involvement in work spills over somewhat to other activities. People's feelings that they are doing their best at work, working their hardest, unaware of time, that they would go on working even if they had no financial reason for doing so, and that they would choose their own job even if they were completely free to make their ideal choice—such feelings and beliefs are positively correlated with membership, activity, and involvement in nonwork organizations (Staines and Pagnucco, 1977). The correlations are low (.25 or less), however, so that the appropriate interpretation seems to be a combination of compartmentalization and consistency, spillover and separation.

Some earlier research also emphasized consistency or spillover from work involvement to nonwork activities. The high level of trade union activity among printers in the United States was so interpreted (Lipset, Trow, and Coleman, 1956), for example. And in Finland, workers who were most involved in their work—eager to improve their occupational skills, apt to make suggestions, and unwilling to change factories—were also most active in the trade unions (Seppanen, 1958; Allardt, 1976).

Research evidence to the contrary is meager, although the dominant assumption has been that workers who were not much involved in their jobs would be somehow compensating through intense involvements of other kinds. Many otherwise valuable research investigations ask people to choose between work and nonwork involvements, thus incorporating the complementarity pattern into the measurement process itself and making it impossible to test for it (Dubin, 1976; Haavio-Mannila, 1971).

In short, what involvement the worker fails to get from work is not therefore likely to be realized in nonwork activities. With respect to involvement, the two sectors of life are relatively independent; to the extent that they are related, the pattern is consistency rather than compensation.

Subjective Reaction

The most subjective relationships between work and nonwork lie in the realm of feelings. Are workers who like their work more or less inclined to like their leisure? Do satisfaction and dissatisfaction tend to be balanced in individual lives, or does one or the other pervade all domains of a person's life?

Once more, the tendency is toward consistency, modest in strength but unambiguous. Detroit factory workers who were satisfied with their jobs were more satisfied with their leisure (Kornhauser, 1965), and the pattern holds for other places and occupations as well (Campbell, Converse, and Rodgers, 1976). Several studies show work–nonwork connections of more specific kinds—for example, between job satisfaction and marital satisfaction. The correlations are strongest in the social aspects of work and nonwork; satisfaction with co-workers is linked to satisfaction with social life off the job.

Many explanations can be proposed for these feelings of consistency. They may reflect some persistent tendency in the method of measurement, some underlying traits of individual personality, or even the effect of nonwork life on life at work. Evidence of consistency is thus not the same as evidence for the effect of the job on other areas of life.

My general conclusion, however, is against the reassuring hypotheses of complementarity and compensation. People seem to have some success in compartmentalizing, in disconnecting their work from their nonwork lives, at what cost we do not know. The tendency toward connectedness comes through nevertheless; the evidence is in the direction of consistency. Those whose advice to men and women doing boring, dirty, or damaging work is "live with it" should not comfort themselves with the idea that these workers realize in leisure the fulfillment that is denied them at work.

MOVING PEOPLE

Leaders and managers of organizations have always been concerned with goodness of fit between people and jobs, but the managerial test of goodness of fit has usually been the ability of workers to do the work. Doing it without harm or boredom, let alone finding joy or fulfillment in doing it, have rarely been among the criteria for selection and placement. Selection, placement, transfer, promotion, and termination are among the traditional functions of management, and all of them are intended to increase or maintain the effectiveness of the organization. They do so by improving the match of people to jobs, with performance as the prime indicator of a satisfactory match.

These people-moving strategies differ from other approaches to improving goodness of fit in that they attempt to change neither the person nor the position. The job specifications remain unaltered, and no effort to educate the individual is implied. The logic is simply to improve the match by rearranging the players.

This does not make the processes either simple or benign. The man or woman who needs a job and is refused on the basis of selection criteria, or who needs to stay on the job and is ordered out, is almost certain to resent the immediate rebuff and may do worse rather than better at the next place. Moreover, goodness of fit is neither one-dimensional nor completely objective. A person may like a job without being an outstanding performer and may perform well at a job that is frustratingly below his or her abilities. (Many men have benefited from the job performance of overqualified women.)

The processes of selection, placement, and termination constitute the dynamics of the labor market as seen from one point of view—that of management. Complementary processes of choice also go on, however. People choose among jobs, decide whether or not to accept transfers, apply for or decline promotions, remain or leave in favor of something more attractive. In simplistic theories of the labor market, these recurring choices of employers and employees are perfectly informed and unconstrained. In practice they are nothing of the sort. Labor agreements, principles of seniority, policies of affirmative action, and other factors modify the choices made by employers; pension entitlements, community ties, peer group friendships, and, above all, lack of opportunity modify the choices made by workers.

There is another sense in which conventional selection and placement strategies are limited: They are inherently conservative. Tests are developed or chosen to select people for existing jobs in existing organizations. To the extent that such tests are validated at all in organizational life, their validity is judged by the performance of men and women in those jobs, rated by their success in those organizations as they stand. It does not exaggerate the case beyond recognition to say that by using such tests meticulously an organization comes to resemble itself ever more closely.

It is possible of course to choose a different criterion for test development and thus to select and promote a cohort of organization changers rather than conservers. Documented instances of selection and placement tests in the service of organizational change are almost nonexistent, but the principle of selection for change is observable in human affairs nevertheless. Especially in selecting top executives and the electing of presidents, organizational or national, the choice may reflect dissatisfaction and the hope for change. Wise choices at high levels sometimes fulfill such hopes, not so much because the selection process itself constitutes organizational change as because it leads to change; the task of change has been delegated to the leaders and awaits them.

But perhaps the greatest limitation to the strategy of improving goodness of fit by "people moving" is that, in a sense, nothing is changed. Jobs are no better, and the capacities and tolerances of people are not directly improved. To the extent that the overall distribution of job characteristics and human characteristics are congruent, moving people can create matched pairs of individuals and jobs. To the extent that the labor market is heavy with jobs that no one finds satisfying, moving people may minimize the residual dissatisfaction but cannot eliminate it.

CHANGING PEOPLE

If goodness of fit between people and jobs cannot be attained merely by making better choices, something else must be changed: the abilities and aspirations of people or the demands and opportunities of jobs. Traditional responses of management have favored the first of these alternatives, and employers spend billions of dollars each year in efforts to change people. These take many forms but we can distinguish a few major approaches in work settings: information giving, training, counseling, *T*-groups, and, more recently, behavior modification.

Giving Information

There is a quality of directness about information-giving approaches to changing people. If workers seem to lack the motivation their jobs require, or are little concerned with the quality of their work, or show only small commitment to the organization, the impulse of leaders to tell them what they should do or be is almost irresistible. Such messages are given in profusion, by word of mouth, in corporate house organs, on factory bulletin boards, and on closed-circuit television. Few such efforts are subjected to serious evaluation, and few would survive it.

The basic flaw in most of these programs is their implicit assumption that a lack of information is impeding the change they are designed to bring about. Where lack of information is the problem, giving information is indeed the solution, but the poor fit of workers to jobs is seldom caused by lack of information. It is not news to the worker that management prefers diligent to desultory behavior, punctuality to tardiness, and commitment to indifference.

A second problem with most informational approaches to improving goodness of fit is the one-way flow of information. The statements are usually made by someone well up in the hierarchy of management, and the flow is downward. The impetus for communication is typically some official dissatisfaction with indicators of organizational performance, and the process of communication is considered complete when the speech has been given or the memorandum distributed. Research on communication, however, demonstrates that a communication cycle is complete only when the intended recipient indicates that the message has been received and understood. And that is a minimum requirement; the intention and the ability to comply are separate issues.

In short, giving information as practiced in most organizations does little to improve goodness of fit between workers and jobs. The demand for adjustment is put on workers rather than on management; the required change is based on managerial criteria of fit; the information flows only one way, and the information content tells workers things they already know about managerial preferences.

Training

Training is a combination of information giving and skill practice, designed to create or enhance the worker's ability to perform some specific task. To the extent that a training program is successful in achieving that immediate objective, it may also improve goodness of fit between workers and their work. It may do so either by making workers' skills and abilities adequate to jobs they already hold or by preparing workers for jobs that will meet their needs more fully.

And there lies a problem. The enhancement of one's skills and abilities is almost always rewarding in itself, but for training to improve the overall goodness of fit between jobs and workers requires appropriate jobs for those newly trained. Indeed, to create qualifications and expectations that cannot be fulfilled reduces rather than enhances goodness of fit.

Self-actualization has two major aspects: the development of latent skills and abilities, and the utilization of valued skills and abilities already developed. Training speaks to the first of these—development; the second, utilization, requires opportunity. If opportunities already exist in an organization, or if turnover and organizational growth will soon create them, training can make a real contribution to goodness of fit. If not, the trained people must put up with jobs that now fit them even less well than before, seek opportunities elsewhere, or change the opportunity structure of the organization in ways that go beyond training.

Counseling

Counseling and psychotherapy are modes of individual change that go deeper, at least in intent, than imparting information or conducting training programs. The target becomes the individual's motivating needs and values, of which he or she is only partly aware. As with other approaches to improving goodness of fit through individual change, two questions must be answered: Does this experience really change people? And if so, are the resulting changes likely to improve the fit between people and jobs? In the case of therapeutic approaches, the answers to those two questions are, respectively; "sometimes" and "seldom."

The assumption that insights gained through therapy can alter deep-seated attributes of personality is supported by a great deal of individual testimony and a smaller amount of evaluative research. In addition, some classic experiments in attitude change (Katz, Sarnoff, and McClintock, 1956) confirm the powerful effect of giving people new insights into their own motives. These experiments demonstrated that prejudices toward blacks could be changed momentarily by information, but that more lasting change resulted from giving people insight into their own motivation for prejudice. Even enthusiasts for one or another therapeutic approach, however, do not expect it to be invariably productive of new insight and personal growth. The answer of "sometimes" to the question of whether therapy changes people simply acknowledges the great variations in the soundness of therapeutic theories, the talents of therapists, and the resistance of clients.

The second question, whether the changes induced by counseling and psychotherapy are likely to improve the goodness of fit between people and jobs, is deceptively simple. Let us assume that successful therapy puts people more fully in touch with themselves and leaves them better able to interpret their own behavior and that of others, less driven by unrecognized motives, and more aware of their own needs and aspirations. None of these outcomes, desirable as they are for the individuals concerned, will make them more satisfied with dull jobs, more docile in response to supervisory pettiness, or more accepting of avoidable hazards. The contributions of counseling and psychotherapy to goodness of fit between workers and jobs, in other words, will be positive in those cases where the worker's neurotic problems are preventing him or her from meeting the expectations and utilizing the opportunities of the job. But such situations may be the exception rather than the rule. I believe that most job dissatisfaction stems from limitations of jobs rather than neuroses of workers and conclude that counseling and psychotherapy seldom improve the goodness of fit between workers and jobs.

The exceptions, however, are numerous enough to deserve mention. People who suffer from serious neuroses, almost inevitably expressed in relation to the job, may be much helped by counseling or therapy. The paranoiac who is too fearful and suspicious to engage in the normal exchange and task-oriented cooperation of organizational life, for example, after therapy may be better able to relate to others and better integrated into the work group.

Moreover, people in positions of substantial autonomy and power may, after therapy, alter their behavior in ways that fit their own altered personalities and at the same time improve the jobs of their subordinates. Menzies (1960) and Sofer (1972) report such "emotional ground-clearing" in a hospital, a technical college, and an industry. They emphasize therapy in organizational settings as the first step in a two-stage process in which the insights of therapy provide the basis for structural alterations in the organization itself.

Most managers, to judge by their actions. have reached similar conclusions about the potentialities and limitations of psychotherapy in work settings. Perhaps the most ambitious program of employee counseling on record was established in the Western Electric Company at its Hawthorne plant more than a generation ago. The counseling program was an outgrowth of the famous Hawthorne Studies (Roethlisberger and Dickson, 1939) and reflected the conviction of management that those studies showed workers' preoccupations with their personal problems to be interfering with their satisfaction and productivity at work. The number of counselors at the plant, which was five in 1936, had expanded to 55 by 1948. But by 1955 it was down to eight, and efforts to introduce similar counseling programs at other locations were rejected. Johnson (1975), in describing these developments, notes that the counseling program at Western Electric declined when a new generation of managers began to ask questions about justifying its costs. He concludes that counseling programs reflect managerial concern for employee welfare, but do not change organizations.

Behavior Modification

The work of B. F. Skinner (1967, 1974) in operant conditioning, known more widely as behavior modification, is based on the principle of "reinforcing" by means of rewards those behaviors that the experimenter desires the experimental animal to perform. Originally developed in the laboratory, with rats and pigeons as the experimental subjects, behavior modification has now been brought into human organizations. The purpose of these applications has usually been to increase productivity, but there are some impressive successes of other kinds, including retraining delinquent boys (Cohen and Filipczak, 1971).

Seen from the perspective of goodness of fit between workers and jobs, these applications of behavior modification are problematic. The problems are not with the psychological or material rewards themselves. The rewards are real, predictable, and valued by those to whom they are offered; the success of the program depends on these attributes. The problems begin with decisions about the schedule of reinforcing rewards. What behaviors are to be rewarded, how should they be rewarded, and should "negative rewards" (punishment) be allowed to enter into the program? In most organizations where behavior modification has been used, these questions have been answered by management. The imposition of a carefully constructed program of behavior modification without the active participation of the men and women who will be affected by it reduces rather than increases employee autonomy and in this respect reduces also the goodness of fit between workers and jobs.

Questions about who determines the rewards of successful effort and the contingent behaviors to be rewarded do not enter into the animal experiments of operant conditioning. The wish to name the game, to choose and define it rather than merely to play it, is distinctively human, and the trend in human organizations is toward recognition of that fact. Appliers of behavior modification to organizations seem not to have recognized it, and they have been criticized for failing to do so (Argyris, 1971;

Whyte, 1972). Indeed, behavior modification has been called a new Taylorism, the successor to Frederick Taylor's (1911) system of piece rates and productivity incentives. Whether or not it deserves that criticism, behavior modification in organizations has done little thus far to improve the match between individual needs and organizational offerings.

T-Groups and Other Encounters

Conventional information and training programs, counseling, and behavior modification, much as they differ one from another, have something in common: all attempt to change individuals as individuals. None of them takes into account the fact that most work is done in group settings and that work groups exert a great deal of influence over their members. Other approaches to change, however, begin with the power of the group, and the best known of such approaches is that of the *T*-group.

Also known as the laboratory method and as sensitivity training, the *T*-group approach can be thought of as a technology of the peer group. A program of such training usually consists of 15 to 20 hours of group meetings, spaced over several days or a week and conducted under the guidance of a professional trainer. The trainer does not provide an agenda, however, or leadership of any usual kind. The theory of sensitivity training, and of its various group-encounter spinoffs, is that as group members struggle to compensate for the absence of traditional leadership, they can be helped to observe the struggle in which they themselves are engaged and thus to become more sensitive to their own behavior, more accurate in their observations of others, and more capable of exchanging information about such matters.

Members of sensitivity groups, who usually come together as strangers for the purpose of learning and exploration, often develop feelings of deep attachment and understanding in their small temporary society. It is difficult to find evidence, however, for improved goodness of fit between *T*-group alumni and the jobs to which they return. Nor have proponents of such group experiences based their case on increased comfort and satisfaction with existing jobs; they have argued that the group experience motivates people to change the organizations in which they work rather than accept them as they are. Some such changes have been widely demonstrated, but they must be judged modest. More than 10 million people have experienced some variant of sensitivity training, and they have made no conspicuous changes in the organization of work.

THE LOCUS OF CHANGE

The strategies of people moving and people changing have a common and crucial element: the locus of change. All of them attempt to improve the goodness of fit between people and jobs by acting upon people in some way—hiring or firing them, moving them from one job to another, instructing, counseling, or sending them off for group experiences in sensitivity training. The underlying idea in all these activities is adaptation, making individuals more suitable for or accepting of the jobs they have.

This idea is implicit in the people-moving strategies. Workers are hired to perform jobs that exist, not to create alternatives to them. People are transferred or fired for failure to perform their jobs, not for failure to change them. And people are promoted, by and large, for performing successfully in the organization as it stands. Exceptions to these generalizations can be called to mind, of course. New executives are sometimes chosen for change rather than continuity, and staff members in research and development are expected to generate ideas for changing organizational products or methods of production. But the generalization holds: Selection and placement, and the less sophisticated procedures of transfer and termination, take the organization as it is and search for people who can successfully do the same.

In the people-changing strategies, the adaptive emphasis is

almost equally clear. Information programs are designed to help people get along in the organization as it is and to persuade them that its well-being is closely linked to their own. Training programs are designed to increase the competence of workers, either for jobs they have or for jobs to which they aspire. The training function thus contributes importantly to goodness of fit between people and jobs, but it does so wholly by changing the abilities of people rather than the content of jobs.

The target of counseling depends substantially on the counselor. To the extent that the effect of counseling is to put the individual more fully in touch with his or her own motives and aspirations, counseling may generate indirect tendencies toward organizational change. Most counseling programs in work organizations, however, are intended to help people cope with their jobs as they are and to do so without bringing into the work setting the problems that arise outside it.

Behavior modification, as the name reminds us, is a method for altering individual behavior, and in principle any behavior can be selected for reinforcement. One could, for example, devise a behavior-modification program that would reward attempts to introduce organizational change. In organizational practice, however, behavior modification has been used primarily to increase individual productivity on routine jobs.

Like other people-changing strategies, off-the-job programs of sensitivity training and similar group experiences have individual change as their first target—increased awareness of one's own behavior, increased understanding of group development. These group strategies are more explicit than the individual strategies, however, in taking organizational change as a secondary target. The men and women who return from successful T-groups, it is sometimes alleged, become "change agents" in their own organizations, movers and shakers toward a more humane organizational life.

The idea that any of the people-changing strategies may have secondary effects of organizational change is intriguing, but it involves an impressive and discouraging series of assumptions.

These are seldom specified, but include, at the very least, the following assumptions:

1.   Exposure to the program will give the individual new knowledge, skill, or insight.
2.   These attributes will alter the individual's approach to the organization and his or her role within it.
3.   These insights and motivations will persist after the individual leaves the special circumstances in which they were acquired and returns to his or her accustomed place in the organization.
4.   The person will change his or her own role accordingly and will persuade supervisors, peers, and subordinates to accept the changes that confront them as a result.
5.   These co-workers will also be persuaded to make complementary changes in their own expectations and behavior.
6.   These changes will be recognized and incorporated into the organization's policies, authority structure, and division of labor.

In short, the people–moving and people–changing strategies, to the extent that they can be said to acknowledge the task of organizational change at all, do so by delegating it to the individuals moved, trained, counseled, or sensitized. It is seldom a successful delegation, and it is not often intended to be. People-moving and people-changing programs improve goodness of fit by changing people, not jobs or organizations.

SUMMARY

Three general answers are identified as responses to the question of goodness of fit between people and jobs: living with it, changing people, and changing jobs. Advocates of the first often argue that people compensate in leisure activities for the deficiencies and demands of their work. Research evidence on this issue, although it is limited, favors the principle of consistency in work and leisure rather than compensation or complementarity.

The idea of improving goodness of fit by changing people involves two rather different strategies—one concerned with changing the assignment of people to jobs and the other concerned with the induction of change in people themselves. The former emphasizes selection, placement, and termination; the latter emphasizes information programs, training, counseling, T-groups, and behavior modification. Both strategies can improve goodness of fit in some cases, but neither addresses directly the improvement of jobs themselves.

CHAPTER NINE

Changing Organizations: Strategies of Coping

The strategies for improving goodness of fit discussed thus far are one-sided. They note the differences in profile between the characteristics of jobs and those of jobholders, and they propose to bring the two profiles more nearly into congruence by altering one of them: the attributes of jobholders. In this sense both the people-moving and the people-changing approaches, including their group variations, are strategies of adaptation. They accept the environment as it is and propose to change (or exchange) the organism.

Adaptation is the dynamic of biological evolution and the means of surviving in environments that cannot be changed. But human organizations are neither unchangeable environments nor natural biological systems. Organizations are synthetic constructions. People invented them, people constructed them, and people are constantly engaged in modifying them, intentionally and unintentionally.

It follows that the fit between people and jobs can be improved by changing jobs and changing the organizations within which jobs are defined. These organization-changing strategies are complementary to the strategies of changing people; they are newer, less familiar, less developed, often suspect, but full of promise. Organization-changing strategies take many forms, but we can identify three major categories, named for the aspect of organizational structure that each takes as its primary target:

1. Changing the distribution of power.
2. Changing the allocation of rewards.
3. Changing the division of labor.

Together, the three domains identified by these approaches comprise a great deal of organizational life and include most of what makes a job desirable or undesirable. The distribution of power involves the authority structure of the organization, the extent to which members participate in the decisions that affect them, the kind of supervision to which they are subject, and the extent to which they are autonomous or controlled in their own work. The allocation of rewards has to do with the major material returns from work—wages and salaries, benefits, and promotions. The absolute amount of these rewards defines the worker's living standard; their relative amount determines the worker's sense of equity or inequity. Finally, the division of labor within an organization, given the organizational product and overall technology, defines the content of individual tasks. The distribution of power, the allocation of rewards, and the division of labor are thus the major organizational determinants of worker satisfaction, the elements of which are discussed in Chapter 3.

Power, rewards, and the division of labor are interrelated. Complex tasks tend to be better rewarded, and power enables its possessors to alter their own rewards or tasks and those of others. Organizations are systems of interdependent parts, and changes of one sort and in one location, if they persist at all, evoke changes of other kinds and in other places. This interrelatedness does not imply, however, that all strategies of change and points of application are equally productive.

Before adopting any strategy of organizational change for improving goodness of fit between workers and work, we should know the answers to four questions:

1. Is the organizational characteristic proposed for change reliably associated with goodness of fit? For example, is goodness of fit better in organizations where power is widely shared than in organizations where it is narrowly concentrated?

2. If the characteristic is associated with goodness of fit, is there evidence that changing this characteristic in the direction implied by such associations really improves goodness of fit? It is one thing to find associations "in nature" and quite another thing to create them experimentally.

3. If such experimental evidence is favorable, are the proposed changes feasible on a large scale? This question involves the technology of organizational change—costs, availability of expertise, and the like.

4. If the technology is available and affordable, what is the potential diffusion of the change? The tendency toward homeostasis and stability is strong in organizations; many attempts at general change are locally contained, and many changes intended to be permanent do not persist. In most situations, strategies that produce stable changes and encourage diffusion are preferable to those that have more limited or temporary effects; the question is how to attain them.

With these questions in mind, let us examine the evidence for three strategies of organizational change: changing power, changing rewards, and changing tasks.

CHANGING THE DISTRIBUTION OF POWER

Power is basic to organizational life; it is impossible to imagine an organization without power over its members. The exercise of power in any organization is limited, however, in various ways—limited in the actions to which it applies, in the sanctions that back it, in the positions from which it may be exerted, and in the positions to which its directives apply. The acceptance of such stipulations, both by those who wield power and those whose actions are controlled by it, legitimizes its use and creates a structure of authority or legitimate power.

The need for power stems from the essential nature of human organizations and human beings. When people say that they are "going to get organized," they mean that they will act in certain interdependent ways and that each of them will know what to expect of the others. The resulting pattern of behavior *is* the organization; if the pattern changes, the organization has changed, and if enactment of the pattern stops entirely, the organization no longer exists.

Such patterns of predictable and interdependent behavior do not occur spontaneously. Left entirely to themselves, people differ considerably in preferred activities, rates of action, hours of awakening, and the like, much as they differ in height, weight, or ability. Moreover, each individual's preferences vary from time to time; we feel more energetic on some days than on others, for example. All organizations therefore require the reduction of individual variability to some degree. In organizational life the normal bell-shaped curve of individual characteristics is altered to the more sharply peaked J-curve of conformity to organizational requirements (Allport, 1933; Katz and Kahn, 1978), as if the pressures of those requirements were applied to the sides of the curve to reshape it (Figure 9.1).

This conformity to organizational requirements is evoked mainly by the exercise of power and authority. The distribution of power varies in organizations, but it always has two aspects: the extent to which the individual's behavior is controlled at all and the extent to which the individual shares in the decisions by which he or she is controlled. The first is the absolute amount of control to which the individual is subject; the second is the relative amount of control exercised by people at one level in an organization compared to those at other levels.

These two aspects of power or control in organizational life

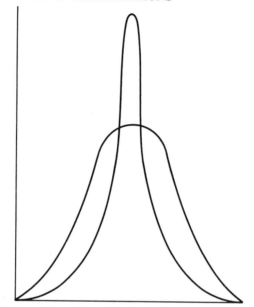

Figure 9.1. Normal, bell-shaped curve and *J*-curve.

can be represented as a single curve (Tannenbaum and Kahn, 1958; Tannenbaum, 1974) in which the slope of the curve indicates the relative amount of control exercised by people at each organizational level compared to those at other levels and the height of the curve represents the extent to which individual members are controlled at all (as compared to being left entirely to their own preferences). The slope of the curve thus distinguishes between hierarchical control, in which the decisions of each level can be countermanded at the next higher level, and democratic control, where the decisions of officers can be overruled by the organizational membership as a whole. The height of the curve, regardless of its slope, distinguishes between strongly controlled and laissez-faire organizations. Figure 9.2 illustrates hierarchical, democratic, and laissez-faire distributions of power in organizations.

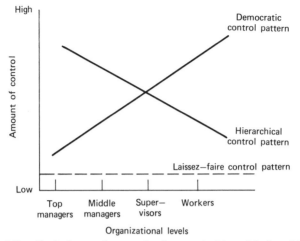

Figure 9.2. Typical control curves for democratic, hierarchical, and laissez-faire organizations.

Power and Goodness of Fit

The hierarchical control pattern is dominant in human organizations generally, and in work organizations it is almost universal. It holds for large plants and small ones, for private enterprise and socialism, and for all industries about which data are available. There are differences, of course, among plants, countries, and types of organizations. Plants in Italy and Austria were found to be more sharply hierarchical than the workers-council plants of Yugoslavia or the kibbutz industries of Israel. Workers in the United States report having more say in decisions than do workers in Italy or Austria, but less than those in Yugoslavian and Israeli (kibbutz) plants (Tannenbaum, 1974). The hierarchical pattern is dominant, however, and it is legitimated.

When workers are asked to describe the pattern of control that they would most prefer at work, their responses still show a hierarchical slope. But—and this is the crucial issue for goodness of fit—the preferred or ideal pattern is less steeply hierarchical than the actual pattern. Rank-and-file members want more say about the decisions that affect them. They do not alway want their supervisors and managers to have less influence in organizational matters, but they want more for themselves. In every country studied, in 73 plants for which control curves were developed, the slope of the ideal curve was more positive than that of the actual (Tannenbaum, 1974). Comparative data on worker satisfaction confirm the message. The needs and aspirations of workers would be better met if their share in decisions were increased.

Experiments with Power

Attempts to change the distribution of power in organizations go by many names, of which participation is probably the most familiar; it is a durable concept in organizational research and a catchword in research application. The underlying assumptions of experimental research on participation have been that rank-and-file involvement in decisions is low in conventional organizations, that increasing it will certainly enhance satisfaction, and may also increase the wisdom of decisions and the motivation to carry them out. Increasing participation thus involves power equalization (Leavitt, 1965), although it may also increase the total amount of power brought to bear on the individual members from all sources, co-workers as well as supervisors.

Many approaches to organizational change include participation and power equalization as secondary or incidental targets. The norms of T-groups and other forms of sensitivity training, for example, emphasize equality of membership, the desirability of expressing one's feelings, and the benefits of consensual decision making. Such ingredients, in contrast to those of everyday organizational life, produce a heady wine but, as we have seen, it does not travel well. To the extent that the qualities of the T-group carry over to organizational life, the effect is to bring the distribution of power nearer to the wishes of most organizational members.

The power-equalizing aspects of survey feedback are more obvious. The process of feedback begins when work groups and their supervisors meet to discuss survey results for their own organizational units and to decide what should be done in response to the findings. To the extent that the processes of information exchange, mutual influence, and joint decision making about such matters "take," the distribution of power has been altered. Research shows that survey feedback does change organizations; the changes are in the direction of power equalization, and they generate concomitant increases in satisfaction (Bowers, 1973; Likert, 1967).

The Clerical Experiment

The distribution of power in organizations can also be increased (or decreased) by direct intervention in the organizational structure, as most managers know and as several ambitious field experiments have shown. One of the earliest of these is known as the Clerical Experiment (Morse and Reimer, 1956), both because its several hundred subjects were office workers and because the company in which the experiment was done preferred not to be identified. Four divisions of the company were involved and in two of them, which formed the autonomous program, the distribution of power was changed so that it was more widely shared. By a decision of top management, rank-and-file employees were given authority and responsibility for functions previously performed by their immediate supervisors, as well as for their own nonsupervisory work. A complementary delegation of authority was introduced at every other level of the company, from first-level supervisors to the executive vice-president.

Many methods were used to create these changes in the distribution of power—persuasion and discussion at the top of the company, review of findings from earlier nonexperimental research, and training of supervisors to help them develop skill and comfort in their new roles. Basic, however, was the executive vice-president's formal introduction of the change as the official policy of the company. Nine months were required for the transition from old to new, and the effects of the experiment were measured one year after the transition was complete. The results, so far as goodness of fit is concerned, were unequivocal. Workers reported increased self-actualization on the job, improved relationships with supervisors, and greater liking for the company. They liked the program, felt that they were gaining much by it, and wanted it to last indefinitely. Productivity increased by about 20 percent.

During the same period, two other divisions of office workers in the company encountered experimental changes of an opposite kind, involving tighter control and increased regulation from above. In these divisions, known as the hierarchical program, feelings of self-actualization decreased, relationships with supervisors deteriorated, and liking for the company dropped. Most workers felt that the company was getting something from the experiment but that they were not; they wanted the program to end immediately. Goodness of fit clearly changed for the worse. Nevertheless, productivity went up by 25 percent, an important reminder that satisfaction and productivity do not necessarily go together.

Management chose to discontinue both programs, in spite of the productivity gains. The hierarchical program seemed too harsh and the autonomous program seened too much like abdicating the conventional powers of management.

The Banner Experiment

The Banner Company is a large manufacturer of packaging materials. At the time of the study, it consisted of two plants, which together employed about 800 workers. The experimental program, initiated because of the vice-president's enthusiasm and conviction, called for increasing the decision-making powers of employees by establishing "overlapping groups" from the bottom of the organizational hierarchy to the top, each group consisting of the men and women who reported to a single supervisor or manager (Likert, 1961; 1967).

The targets were several, as were the methods, but the key elements were worker access to information and involvement in decisions that had previously been reserved for management. The experimental changes were introduced in three departments; two others were treated as controls, and in them no changes were proposed during the experimental period. During the 30 months of the experiment, the three experimental departments showed increases in machine efficiency and in worker responses to a seven-dimension index of overall satisfaction. In the control departments both efficiency and satisfaction decreased somewhat during the same period.

Despite some unexpected findings, such as the failure of the experimental departments to show increases in workers' satisfaction with their work groups, the overall inference seems clear: Workers at Banner, like those in the Clerical Experiment, became more satisfied with their total work situation when they were allowed more of a share in making decisions about things that affected them. In both experiments, modifying the power structure to include more input from employees brought the organization more in line with their needs, and in both there were concomitant gains in productivity.

The Weldon Experiment

The Weldon Experiment in organizational change was built around a company acquisiton, routine in some respects and remarkable in others. Acquisition by purchase is routine, and this was an acquisition of Weldon by its long-time competitor Harwood. The two companies were about the same size, with 1000 employees, and about the same age, 30 years old. They used a common technology and manufactured similar lines of men's wear.

But there the resemblance stopped. Harwood had a 30-year

commitment to participative management; Weldon was no less committed to conventional or even authoritarian managerial practice. The Harwood management accepted and bargained comfortably with the Amalgamated Clothing Workers Union; at Weldon, the same union had just lost an election, after a long and expensive campaign. During the year of acquisition, Harwood showed a 17 percent return on invested capital; Weldon showed a loss of almost the same magnitude. Production at Harwood was 6 percent above time-study base; at Weldon it was 11 percent below. Worker earnings were higher at Harwood; turnover and absence were about half the Weldon rate.

For economic reasons, the Harwood management faced the task of bringing the performance of its former competitor up to its own level. And for reasons of managerial philosophy, the Harwood management chose the authority structure as the primary target of organizational change. A program was developed to move managerial behavior at Weldon from "system 1" toward "system 4," that is, from authoritative toward participative management (Likert, 1961; 1967). Several methods were used to bring about this change: Harwood managers not only espoused participative management, they exemplified it. Social science consultants helped train Weldon managers in participative decision making, initially in off-the-job settings and subsequently through coaching in the work situation. These kinds of training were begun at the top of the company and steadily extended, with each training unit consisting of a manager or supervisor and those employees who reported directly to that person.

Changes in authority patterns and related managerial behavior were measured by questionnaires that emphasized six areas: control, decision making, goal setting, interaction, communication, and motivation. Comparisons of managerial reponses before and after the program show significant change toward participative management in every one of the six areas, for every level of management and, although less sharply, in the reponses of workers.

The outcome suggests clear improvement in goodness of fit.

Workers' attitudes toward their jobs, the compensation system, and the company all became more positive during the two-year period, and workers said that they were putting more effort into the job. Harder evidence consistent with their questionnaire responses showed up in the economic changes. Efficiency as measured by time study increased from -11 to $+14$ percent, and incentive earnings above base went from zero to 16 percent. Turnover and absence, behavioral indicators of disaffection, dropped by half. The rewards of ownership were no less dramatic; return on invested capital rose from -15 to $+17$ percent.

Like most experimental programs in actual organizational settings, the Weldon experiment departs from the textbook prescriptions for experimental design. There was no control group, no second Weldon from which the change program could be withheld. Many specific actions were taken to bring about the desired changes, so that no single experimental stimulus can be identified. Finally, external events may have added their effects to those of the changes in authority structure. Through all these cautionary notes, however, the experimental findings make themselves heard persuasively. At Weldon, as in the Banner and Clerical Experiments, the distribution of power and authority was changed in ways that increased employee participation in decision making, and employees indicated by words and actions that their work better suited their needs, aspirations, and abilities.

Feasibility, Durability, and Diffusion

There are three additional questions that pragmatic leaders must ask of a successful experimental procedure: Is it feasible on a large scale? Will it last? Will it spread? The Weldon, Banner, and Clerical Experiments speak to these questions, although none of the experiments was designed to deal with them explicitly.

The question of large-scale feasibility is perhaps the easiest to answer. All three experiments involved hundreds of employees, and all three showed performance gains that more than compensated for their costs. Moreover, a substantial part of those costs was incurred precisely because the programs were experimental, new, and subjected to considerable observation and measurement. No insurmountable problems of feasibility are apparent in extending such programs, and the managements of many organizations have emulated them to some degree. It could be argued that instant nationwide adoption of participative management would create a demand for expert consultation beyond the available supply, but this is an unlikely turn of events. Moreover, the procedures for altering the distribution of power are not mysterious or beyond the present staff resources of most large organizations.

Evidence for durability is spotty but encouraging. None of these experiments was brief. The first measurements in the Clerical Experiment were taken after the experimental programs had been operating at full scale for a year. The preexperimental and postexperimental measures were about 30 months apart at Banner and about two years apart at Weldon. Some evidence for durability is implied by these facts alone. At Weldon, moreover, a third set of measures was taken five years after the termination of the experiment. On all six of the relevant organizational dimensions (control, decision making, goal setting, communication, interaction, and motivation), management was more participative than it had been at the end of the experiment five years earlier. Not only had the experimental changes endured, they had continued to develop.

These three experiments in the distribution of power illustrate the process and problems of diffusion within the experimental companies and beyond. In the Clerical Experiment, as we have seen, corporate management decided against wider adoption or even continuation. Informal reports from the company reveal that this decision was readily accepted by employees in the hierarchical program and strongly opposed by those in the autonomous program. Interest in other parts of the company ran high, and some potential for diffusion almost certainly existed. It was never realized.

The Banner Experiment showed early signs of diffusion when departments that had not originally been scheduled for the experimental program asked to be included in it. What more might have happened had the company remained independent we cannot know. Banner was acquired by a much larger corporation, with a strong and differently oriented management. The Banner Experiment was not cogenial to the new corporate management, and the diffusion of management practice was toward Banner rather than toward the new parent corporation. The Banner vice-president who had been most involved in the experimental work left the company to become a management consultant specializing in the introduction of participative managerial practices. Thus it might be said that the Banner Experiment generated a force toward diffusion—but in other companies.

It is in other companies, however, that the larger question of diffusion must be answered: Can experiments in power equalization play a significant role as demonstration projects? An affirmative answer will have to wait either for stronger evidence or for a more sanguine observer. Experiments in the enlargement of employee decision making continue, but they have not swept through American management as time and motion study did in an earlier generation or as the use of computers is doing now.

Sharing the prerogatives of conventional management requires managers themselves to change and, in a sense, to change themselves. Adopting a new technology delegates the change demand to other levels of the organization. There is almost certainly some movement toward a more consultative, information-sharing managerial style in the United States, but I would call it a cultural drift rather than anything more rapid or vigorous, and its future is not clear. Experiments in sharing authority have been successful, by and large, but the problem of diffusion both within the experimenting companies and beyond them, has yet to be solved.

CHANGING THE ALLOCATION
OF REWARDS

The intrinsic satisfactions of work and the importance of friendships made at work enlarge upon but do not eliminate the essential instrumental character of work. Work is first of all an instrumental activity, rewarded with money in industrial societies and rewarded in others by barter or a direct share in the common product. In societies that have a money economy, money becomes a nearly universal reinforcer and is so recognized. Labor negotiations are concentrated more on wages than on any other issue, and governments may stand or fall on their policies regarding wages and prices. Every serious study of the relationship between workers and work affirms the importance of pay as a determinant of overall satisfaction.

Pay is important primarily for what it buys, of course—food, shelter, clothing, medical care, recreation, education, and all the other goods and services that make up the gross national product. But the symbolic significance of pay is also extremely important. No other language conveys managerial approval of person or performance so clearly as a significant increase in pay and status. Furthermore, the relative distribution of pay in an organization reveals a great deal about managerial philosophy and practice, and determines workers' feelings of equity or inequity.

Equity in pay is a persistent problem in any organization, because the criteria of equity are many and not wholly compatible. What weight should be accorded to seniority or performance, to skills possessed versus skills actually required on the job; what compensation, if any, should be given for hazard or hardship, what rewards for supervisory and managerial responsibility? And beyond such questions about criteria are the inevitable disagreements about individual ratings on whatever criteria are in force. To attain consensus in this sensitive area is difficult, but serious managerial efforts to do so satisfy workers beyond the purchasing power of the dollars themselves.

Goodness of Fit

Most research on pay has been done in conventional organizations, which have similar reward structures. In such organizations pay and perquisites increase as one moves to jobs of increasing complexity, and compensation increases more sharply as one ascends the hierarchy of management. Pay is usually computed for time worked and, as rewards increase, so do the time units by which rewards are computed—an hour for blue-collar workers, a week or a month for office workers, and a year for executives and highly placed professionals. For people doing identical or similar jobs, relative pay is usually determined by seniority, performance, or both. Most research on rewards in work settings has not involved change and therefore cannot answer directly our questions about the effects of changing the reward structure; it can, however, teach us how the rewards of work meet or fail to meet the needs of workers.

The first lesson of such research is the importance of pay. For 50 years, in countries from the United States to India, researchers have measured the importance of wages as a reason for selecting a job, leaving it, or remaining on it in varying states of satisfaction. By all these criteria, and for all populations—from unskilled laborers to scientists and managers—the research findings are in agreement: Pay is almost always included among the three most important aspects of a job and is often rated first. (Lawler, 1971; Fein, 1976).

People are not naive about work. They distinguish different sources of satisfaction and dissatisfaction and may say, for example, that they are satisfied with their supervisor but dissatisfied with the content of their job. Nevertheless, pay is so important that its effects generalize. The amount of pay in relation to workers' needs and expectations not only determines satisfaction in its own domain, but strongly affects satisfaction with the job as a whole.

The second research lesson on the importance of pay is that

people have different needs and differ therefore in the importance that they attach to pay. Men rate pay more important than women do, according to past research, although that may change in the future. Young workers rate pay more important than older ones. And some fragmentary evidence suggests that personality differences affect the relative importance attached to monetary rewards; people whose self-assurance is low consider pay more important than those who are more self-assured and less anxiety-ridden.

The third lesson from research on the importance of pay is that, as elsewhere, circumstances alter cases. People in high-level jobs say that pay is less important than do people in lower-level jobs, a finding that suggests a hierarchy of needs (Maslow, 1943, 1954) in which the felt importance of a need recedes as that need is fulfilled. Pay is also rated differently by people in different kinds of organizations, with higher ratings coming from people in profit-making organizations than from those in government and social service jobs. Such differences almost certainly reflect both self-selection and acculturation. People who care most about money will seek out organizations that offer the prospect of large rewards, and people working in such organizations will tend to become more reward-conscious.

"Experiments" with Pay

Most workers are paid for units of time, units of production, or some combination of the two, but most say they would like their pay to reflect their performance. Nevertheless, rank-and-file preference the world over has been for hourly rates rather than piece rates, and in the United States hourly rates are the far more common method of pay for industrial and service jobs. Three innovative reward systems have been tried extensively, however, all involving increases at nonsupervisory levels: profit sharing, the Scanlon Plan, and employee ownership (directly or through the purchase of stock). None of these has been given a true experimental evaluation, but all three have been the subject of considerable research, and the results are instructive.

The Profit Sharing Research Foundation has made a number of economic comparisons between profit-sharing and nonprofit-sharing enterprises of similar kinds, and the results favor the former. Moreover, a few profit-sharing plans have attained prolonged and widely publicized success. The Lincoln Electric Company, which manufactures welding equipment and is dominant in its field, has had a profit-sharing plan for more than 40 years and routinely pays in shared profits as much or more than it pays in basic wages (Fein, 1976). What this does for workers' health is not known, but low rates of absence and turnover suggest that it does a good deal for satisfaction. Even in times of labor shortage, there is reputed to be a long list of applicants at the Lincoln Electric Company.

These applicants are not random members of the labor force, of course. The reputation of Lincoln Electric, both for high pay and hard work, is well known in the community, and many workers may not find the bargain attractive. Moreover, the company hires applicants selectively, so that the resulting employee group may be a population of money-oriented work athletes rather than a representative sample of the adult labor force. Furthermore, Lincoln offers advantages in addition to profit sharing—guarantees of job security, options to purchase stock, an elected board of employee advisors to the company, an unusual degree of autonomy at the job level, and a relatively enlarged definition of assembly jobs (Fein, 1976).

The Scanlon Plan can be regarded as a prolonged and continuing company experiment with pay. It includes a productivity-based method for determining worker pay beyond the standard hourly rates, a system of related production committees that begin with elected representatives of the workers in each department and reach to the top of the company, and a continuing role for a labor union as one of the two contracting parties to the overall plan. Joseph Scanlon, labor organizer turned academic, originally proposed that all direct monetary gains from performance over the preplan level should go to the workers; management and stockholders would benefit only from the indirect gains of more intensive utilization of the plant and equipment. Most Scanlon Plan companies—almost 700—use some variation of the original plan and reach their own agreement on the exact basis for sharing revenues.

The Scanlon Plan works. Many companies use it, and workers receive at least 25 percent above the union contract. There have been failures over the 40 years of the plan's operation, but recent reviews (Frost, Wakely, and Rhu, 1974) attribute these to managerial unwillingness to implement the full specifications of the plan itself, especially its participative committee structure. Productivity increases in Scanlon firms run 20 to 25 percent for the first year of the plan and about as much again for the second year. The Scanlon Plan clearly involves a strong economic incentive.

It can be argued that the ultimate financial incentive is ownership, and a small proportion of business establishments offer that incentive. A recent search (Conte and Tannenbaum, 1977) discovered 472 firms that had adopted plans for partial or total employee ownership, and it is likely that an equal number was not turned up in the search. A comparison of 30 such firms with conventionally owned firms in the same industries showed the employee-owned firms to be more profitable. Moreover, the single most important correlate of profitability is the percentage of equity owned by rank-and-file employees. The larger the proportion of employee ownership, the more profitable was the firm (Conte and Tannenbaum, 1977). Case study reports indicate that both workers and managers consider the work situation to have improved as a consequence of employee ownership of stock—better relations among people, better benefits, and more interest in work and in the company.

Feasibility, Durability, and Diffusion

Experience with profit sharing, Scanlon Plans, and employee-stock-ownership plans is sufficiently long and broad to answer the question of feasibility almost unequivocally: The plans work, and the procedures for introducing them are clear enough to permit their use on a larger scale than has yet been attempted. The only reservation has to do with company size. Most of the firms that have tried these modifications in the reward structure are small, and none is among the corporate giants. I believe that these plans can be adapted to large enterprises, but it would be necessary to make the rewards responsive to the performance of visible local groups rather than dependent on striking some distant corporate balance.

Questions of durability have also been answered over the years. Many companies have had profit sharing or Scanlon Plans for decades, and the abandonment of such plans seems to have been determined by changes in managerial philosophy rather than direct indications of failure. Whether the employee-stock-ownership plans are equally durable remains to be discovered. Some employee-owned plywood factories are more than 50 years old. Many more recent conversions to employee ownership, however, have been last-resort, job-salvaging efforts in companies (or divisions of companies) that are threatened with failure or abandonment by the corporate owner. The durability of employee-owned companies is being tested, therefore, under peculiarly difficult conditions.

Nevertheless, both questions of feasibility and durability can be answered positively, if cautiously. It is the question of diffusion that must be answered negatively. Two aspects of diffusion are worth distinguishing, a qualitative and a quantitative aspect. By qualitative diffusion of an organizational change, I mean its spread to other subsystems of the organization. For example, does a change in the reward structure of a company (profit sharing, Scanlon Plan, employee stock ownership) lead to spontaneous complementary changes in the distribution of power or the division of labor in that company? The answer is no. Some profit-sharing plans, like the Scanlon Plan, incorporate formal changes in the decision-making structure of the company. But there is no evidence that, without such provisions, changes in the reward system extend naturally into other organizational systems. Most profit-sharing companies are conventionally managed in other respects. The same can be said of most companies with plans for employee stock ownership.

It is the quantitative aspect of diffusion, the spread of an innovation from one organization to another, that is usually intended when the word is used, however. We must ask whether the adoption of profit-sharing schemes by some companies tends to spread to others, and the answer must be generally negative. After 40 years of profit-sharing plans, Scanlon Plans, and variants thereof, they remain a minority phenomenon in the United States. The performance of such companies, as the Profit Sharing Research Foundation points out (Metzger, 1966), is significantly and persistently above that of their competitors, but neither the precept nor the example has moved the majority of companies to emulation. Moreover, the adoptions have been fewest among large companies, so that the proportion of the non-managerial labor force involved in corporate profit sharing is very small.

The effects of such programs on health have not been examined. I would predict that these plans have positive effects, partly because of increased satisfaction with the work situation and partly because of the improved access to goods and services that significant income gains make possible.

CHANGING THE DIVISION OF LABOR

Human organizations have sometimes been defined as inventions for accomplishing tasks that cannot be done by individuals working alone. Whether or not we agree with that definition, it is useful to think of an organization as engaged in some overall task—providing a service, moving something, or making a product. And whatever the overall organizational task, it must be divided into chunks that can be handled by individuals. The division can be made in many ways, depending partly on the tools and technology to be used and partly on the social arrangements for using them. Nothing about an organization is more basic than this division of labor; it determines what people actually do on their jobs, whether they work with others or in isolation, whether their individual task is varied or monotonous, and whether they are constrained or free to move about.

One way to change organizations, therefore, is to begin with the division of labor itself and thus change the tasks that individuals perform. This is the key idea of what has come to be called the sociotechnical approach to organizational change (Trist, Higgin, et al., 1963; Rice, 1958, 1963; Trist, 1976). The term reminds us that human organizations are sociotechnical systems, always incorporating both a technology (tools and rules for their use) and a set of social arrangements (a structure of power and authority, a pattern of reward allocation, etc.).

The sociotechnical approach to organizational change is concerned with improving goodness of fit in two respects—goodness of fit between the social and the technical aspects of the organization itself, and goodness of fit between the sociotechnical properties of the organization and the needs and abilities of its individual members. Like the other approaches to organizational changes we have discussed—power equalization and profit sharing—sociotechnical change can be undertaken on a modest or an ambitious scale. It may be directed to changing only one set of jobs, it may attempt to change the functions and jobs of whole work groups, or it may alter the division of labor throughout the entire organization. Similarly, the sociotechnical approach may in a particular instance limit itself to changes in either the social arrangements or the technical arrangements, or it may attempt to deal simultaneously with both.

Table 9.1 sets out these six broad possibilities for organizational change as defined by the sociotechnical framework.

Table 9.1   Possibilities for Organizational Change

	System Aspects	
Organizational Units	Social	Technical
Job (role)	1	4
Work group	2	5
Organization	3	6

The diagram reminds us that the approaches to organizational change discussed earlier—power equalization and profit sharing—are concerned primarily with the social aspects of organization. Neither approach (the Scanlon Plan excepted) deals explicitly with organizational technology; that was accepted as given.

The early sociotechnical experiments, despite the implied promise of the name, also accepted the technological aspects of organizations as given. The experimenters believed that many different social structures could be utilized for each given technology. The technology set limits on the possible social arrangements but did not prescribe more specifically. The challenge was to create for each technology a social organization that would both accommodate the technological requirements and meet the needs of the organization's members.

One of the earliest of the sociotechnical studies contrasted two organizational arrangements for using the recently introduced face-conveyor and other then new machinery for mining coal in England (Trist and Bamforth, 1959). The conventional long-wall organization, imposed with the introduction of this machinery, was an obvious imitation of conventional mass-production manufacturing—specialization by shift and by job, so that each miner performed only one task and was paid according to the task. The composite long-wall system developed in a few mines as an alternative social organization; it used the new technology but preserved some of the valued social characteristics of earlier mining methods, in which each miner exercised all the basic skills and was part of a group that was responsible for the complete task of cutting, loading, and doing the necessary construction to advance the coal face and safeguard the miners. Pay within the work group was equal, with a common base rate and a productivity incentive based on the performance of the group. The composite organization outperformed the conventional organization dramatically, with more productivity, many fewer absences, and less supervision.

Another of the early sociotechnical studies involved two textile mills in India, in which the introduction of automatic looms had failed to provide quality and productivity in any way superior to the manually operated looms they replaced (Rice, 1958). The social organization that had been created for tending the 224 automatic looms involved 12 specialized roles, filled by 29 workers. No worker or group of workers within the 29 worked together on a continuing basis, none completed any meaningful unit of work, and none controlled the pace or content of his or her task. The sociotechnical solution was to reorganize the 29 workers into small groups of workers (about seven workers per group) and to make each group responsible for specific looms. These groups took over the ancillary functions that had been assigned to specialists. The results were rapid and dramatic. Productivity during the two-year period after these changes averaged 95 percent of the estimated potential of the looms, compared to 80 percent prior to the changes. Quality defects dropped steadily, from 32 percent in the preexperimental period to 15 percent for final six months of the experiment.

Feasibility, Durability, and Diffusion

In the decades since these early sociotechnical experiments by Rice, Trist, and other members of the Tavistock group in Britain, the number of such studies has multiplied and many countries have developed their own variants. This aggregation of experience provides some information about the questions of feasibility, durability, and diffusion.

As with the experiments in power equalization and reward allocation, the feasibility of organizational change seems clear; the durability of such change seems equally clear for periods of one or two years but less clear for the longer term unless the changes are organizationwide. And the question of diffusion, throughout the host organization and across its boundaries, is least clear of all. Enough is known to design and redesign human organizations in ways that better fit the needs of their members. The process of organizational change is not easy and the results are not assured, as a recent book of documented failures (Mirvis and Berg, 1977) reminds us, but the message of the organization-changing experiments is encouraging: Human organizations can be made more congruent with human health and well-being. Moreover, there are signs that the scattered theories and experiments of the past are beginning to converge.

SUMMARY

Three strategies for organizational change are considered in this chapter, each identified by the aspect of organizational structure with which the change process is begun: the distribution of power, the allocation of rewards, and the division of labor. Each of these strategies is evaluated by reviewing the research evidence, considering the feasibility of the strategy for organizational improvement on a large scale, and predicting the prospects for diffusion of such improvements. All three strategies have had their experimental successes, and some versions of them—profit sharing and the Scanlon Plan, for example—are operating successfully in many companies. None of these approaches has become dominant, however, and there is some reason to think that they may be more effective in combination than in more specialized form.

CHAPTER TEN

Convergences and Combinations

The few classic experiments in organizational change presented here and in earlier chapters should not be interpreted as a representative sample. Scientists who study human organizations, like their colleagues in other fields of research, write books and articles about the experiments that work. They may learn from their failures, but they publish their successes.

Only one book (Mirvis and Berg, 1977) is devoted entirely to the chronicle of failed experiments in organizational change, and it serves to remind us of the many ways in which failure occurs. Some would-be experimenters never gained real entry to the organizations of their choice; management, unions, and sometimes both rejected either the experimental proposition or the experimenter. For other experimenters in organizational change, failure came in attempting to alter the hypothetical "experimental variable"; they had identified some factor they thought would make a difference, but they lacked either the skills or the resources to alter it. In some cases the experimental "cause" was indeed altered according to plan, but without the hoped-for effect. Finally, there are those organizational experiments that, although briefly successful in one place, failed in a larger sense; they neither endured nor spread.

165

No one knows the failure rate of research attempts at organizational change, but the many ways by which such experiments can fail suggest that, for scientists as for managers, the rate is high.

RESISTANCE TO CHANGE

Failures in organizational change call our attention to characteristics of organizations as well as of experiments. To be organized means to have established ways of doing things and to maintain those ways over considerable periods of time. An organization that changed in response to every internal or external murmur would be no organization at all. It would have lost the stability-maintaining properties necessary for its survival.

The problem is that the same properties that enable an organization to survive and keep on course under changing or even turbulent conditions also make it resistant to change in general, perhaps especially to the kinds of structural change in power or rewards or division of labor that would have the broadest effects on work and health.

Some of these resistant properties are obvious. If workers and work groups are to become more autonomous and participate more fully in decisions that affect them, managers and supervisors may fear corresponding reductions in their own authority. Similar resistance is likely to emerge to proposed changes in the structure of rewards, with some groups in the organization fearing that more generous allocations to others must surely reduce their own share of the earnings. And changes in the division of labor, to the extent that they enlarge and enrich previously routine tasks, may threaten the expertise of specialized groups.

These sources of opposition to organizational change share an underlying assumption that is more often false than true—the assumption that organizational life is a zero-sum game. In the perfect zero-sum game, which has provided the framework for many laboratory experiments, the resources and rewards are fixed in quantity. Nothing that players can do will increase that amount. The only possible exercise of skill, therefore, is to increase one's own share of the fixed total and in doing so inevitably reduce the share of the other player. It is an adversary situation.

Organizational life, fortunately, is more complex and often includes the possibility that all, or almost all, players might benefit from the same change. Power, for example, is not a zero-sum phenomenon in organizational life; one may gain and keep the ability to influence another person by being responsive to the requests of that person. In such a reciprocal relationship each has acquired power. Research in many countries and in many kinds of organizations (Tannenbaum, 1974; Tannenbaum and Cooke, 1978) has shown that such reciprocal processes can increase the total amount of influence in an organization and thus contribute to its overall effectiveness. Such gains in turn tend to increase the resources that the organization acquires and therefore the rewards available for allocation. The process of allocation still may be hotly contested, but the constraints of the zero-sum game have been broken; there can be more for all.

Other sources of resistance to organizational change, however, are inherent in all organizations. Their stability is, in the language of systems theorists, overdetermined; that is, there is more than one stability-producing mechanism. People are selected to fit the requirements of the organization as it is at the time of their selection, not as it might be at some later time. Having been selected to meet present requirements, they are trained and rewarded for fulfilling those requirements and perhaps punished for failing to do so. The demand to change thus must overcome the success of previous demands and also the inertia that makes all of us persist in old habits that have become known, familiar, and perhaps easy.

Finally, there is the resistance to change that stems from the necessary interdependence of organizational parts. Change must begin somewhere, not everywhere simultaneously, and change in one part of an organization creates demands for change in other parts. Those demands may be resisted for any of the reasons we have described, and the effects of such resistance are often to weaken or reverse local changes.

DESIGNS FOR FAILURE

The properties of organizations that make them resistant to change are instructive for the design of experiments in organizational improvement. They imply, among other things, that the experimenters had better "start strong," take full account of the interdependence of organizational parts, build in sufficient incentives for change, provide reassurance against possible risks and losses, involve the people who will be affected in the decisions to be made, arrange extensive coaching and instruction in the new ways of doing things, and maintain considerable flexibility about the pace and method as the process of change proceeds.

All this may sound quite unremarkable and perhaps no more than common sense might suggest, but it runs counter to established doctrine on the design of scientific experiments. Textbook descriptions of the ideal experiment, usually presented with examples from the physical sciences, run something along the following lines. The experimenter decides on a hypothesis to be tested and devises a test that involves changing a single variable—increasing pressure or temperature, for example, or adding some catalytic element to a chemical reaction. All else is controlled and remains unchanged. Everything has been planned in advance; once the experiment has begun, no changes may be made in the plan. Since the elements in the experiment are inanimate, questions of their participation in planning or modifying the experiment do not arise; they do not require instruction or assistance in responding to the new experimental situation, nor do they resist its newness in preference for some previous familiar state. The experimenter is in sole charge of the experiment.

Some social scientists have begun to realize that this model, powerful as it has been for the physical sciences, is inappropriate for large-scale experiments to test social policies or organizational designs. Nobel laureate Herbert Simon (1974), arguing for increased breadth in experimental designs, points out that the orthodox approach to experimentation, with its insistence on testing a single effect of a single "cause," produces at most a "one-bit" yield of information—that there is a relationship between an independent and a dependent variable. And in those cases where a combination of causes is required to generate the effect in question, an infinite series of one-cause-at-a-time experiments will yield an infinite series of negative findings. Simon proposes that experiments should go beyond merely demonstrating that the presence of an experimental variable makes or fails to make a "significant difference."

James Morgan (1981) is especially critical of the rigidity of orthodox experimental designs. Reviewing the ambitious economic experiments with negative income tax payments to families with substandard incomes, he emphasizes the importance of flexibility and continuing feedback in such designs. He proposes that continuing information about the progress of such complex and long-lasting experiments while they are in progress is important for their success. That information would be used to modify and adapt the experiment to deal with emerging inadequacies or misunderstandings. The continuous monitoring and modification would thus become a true cybernetic process of self-correction to keep the experiment on course. Careful measurement of the modified experimental inputs and the successive responses to them would retain the scientific value of the experiment and permit its replication. The undeniable losses in experimental control would be more than offset by gains in experimental success. And although one can learn from failure, that learning must ultimately be demonstrated by success.

Chris Argyris (1968) makes an additional and telling point about experiments in organizational settings. He describes in detail the disturbing resemblance of the experimenter's traditional role to that of the authoritarian manager. Both define the task of the subordinate (experimental subject or employee) clearly and rationally, but without input from the subordinate. Both control the pace and method of work. Neither the experimenter nor the authoritarian manager shares information with the subordinate about the larger purpose of the enterprise; the subordinate is given only the information necessary to perform the assigned task. Both the experimenter and the manager decide unilaterally the rewards and penalties of task performance. Both, in short, create a world in which the behavior of the subordinate is "defined, controlled, evaluated, manipulated, and reported" (p. 291).

The evasive and resistant behavior of employees to such treatment has been well documented, as Argyris points out. It includes physical withdrawal (absence and turnover), psychological withdrawal or lack of effort, overt and covert hostility toward those in authority, increased emphasis on monetary rewards, and development of counterorganizations such as unions. The predicted behavior of experimental subjects under comparable conditions would be similar. The effect on the experiment would be to increase the prospects for failure and to present the experimenter with findings in which the intended effect of the experimental treatment is confounded with the unintended effects of experimental "rigor."

For all these reasons, I call the traditional experimental procedures, especially when they are attempted in organizational settings where adult men and women are "playing for keeps," designs for failure.

SIGNS OF SUCCESS

Although failures are common in organizational change, both because of the resistance of organizations and the limitations of orthodox experimental designs, there are signs of success. Moreover, as the successes become more numerous and their research documentation more complete, the requirements for success become increasingly clear.

Experiments that have taken as their primary target some major aspect of organizational structure—power and authority, rewards, or division of labor, for example—account for more than their share of the successes. Such experiments tend to use several methods for inducing change, in combination or in sequence. These include formal managerial decision, group discussion, instruction and training, regular feedback about progress toward agreed-upon goals, and many others. These experiments illustrate the interdependence of organizational structures. Changes in the division of labor, for example, tend to raise questions about changing the allocation of rewards, and changes in the distribution of authority legitimize the discussion of altering rewards, the division of labor, and technology. The prospects for success are greatest, I believe, when plans for organizational change take such interdependencies into acccount.

Still other markers for success include the scale, duration, and flexibility of the change effort. Experiments in the quality of working life have been conducted on different scales, from a few isolated jobs to an entire plant. If we use "experiment" loosely enough, legislation like the Norwegian Work Environment Act of 1977, which puts self-determination and personal development among the legal entitlements of men and women at work, can be thought of as launching work experiments on a national scale. Much can be learned by well-designed and measured experimentation at any of these levels, but I find most encouraging the experiments in the middle range, which deal with individual plants or departments of substantial size. These are units large enough to teach organizational rather than merely individual lessons, and sufficiently autonomous not to be quickly overwhelmed by the resistance of surrounding parts of the organization.

The duration of field experiments on this scale is long by conventional standards. They are likely to require months to design; they almost certainly take many months to attain stability, and several years may be required to assess their effects. Short-term organizational gains, as every manager knows, can be achieved by many methods, including those that imply losses in the long run. The immediate response to pressure may be productivity, as may the immediate response to forcing machines beyond their rated capacities. The longer-term effects may be turnover and early replacement of equipment. Experiments in organizational change need to run long enough for their costs and benefits to be fully assessed, and that is likely to take several years.

Flexibility is the least documented of these markers for experimental success, because the published accounts of organizational experiments seldom include the daily detail of trial and adaptation that went on within the larger design. Observation of such experiments, however, and discussion with those intimately involved in their conduct argue for the importance of flexibility as a factor in experimental success. No matter how careful the plan and how detailed the design, the process of experimental change raises new questions and problems. To refuse these or ignore them, or to reiterate the original instructions when they are no longer applicable, is to invite failure. The more successful researchers tend instead to respond to such issues within the major aims of the experiment, to record in detail the additions to the experimental treatment thus made, and to take them into account in the continuing measurement of experimental effects.

WORK IN AMERICA

Work in America is the report submitted in 1973 by a special task force to the Secretary of Health, Education and Welfare, as that department was formerly called. It is a statement of advocacy as well as a summary of research, and its arguments for the humanization of work aroused great support and great opposition, both of which seem now to have subsided. Less subject to argument were the results of 34 field experiments in the quality of work life, almost all of them in the United States. The list of these experiments was by no means complete even on the date of its publication, and yet it tells us a great deal. The number of people included in such experiments ranged from 6 to 6000, and the experimental aims varied from specific (reduce absence and turnover) to very general (increase the meaningfulness of work). The methods of change varied no less, from the creation of a new plant and organization to the modest enrichment of a few jobs.

The results of these experiments emphasize two points. First, the experiments work; the quality of working life can be improved and the improvements can be achieved without losses in productivity. Second, the markers for success and failure that we have considered are conspicuous in these studies.

Most of the experiments involved some change in organizational structure—in power, rewards, division of labor, or all three. Changes in the division of labor through job enlargement and enrichment were common and in some cases included changes in technology, as at the Medfield Plant of the Corning Glass Company where the assembly line was abandoned. Increases in the autonomy of individual workers and work groups, especially in determining immediate production goals and methods, occurred in about half of the experiments. Changes in rewards were least common, an exclusion that probably would be challenged by employees and their unions in the long run.

These points are well illustrated by one of the most important and most publicized of the work studies, usually identified by its geographic location as the Topeka Experiment. The Topeka Experiment was initiated in 1971 in a new plant of the General Foods Corporation intended to manufacture dog food. Several years of planning preceded the opening of the plant, which was designed to provide production workers with unusual amounts of responsibility, autonomy, and variety on the job. Since the plant was new, there could be no comparisons of its experimental performance with a conventional preexperimental period. Comparisons were possible, however, with a conventional plant of the same parent company in the same region. Comparisons could also be made between the actual performance of the plant and the costs and output obtained by conventional methods, as estimated by industrial engineers. By both these standards, the Topeka Experiment, some five years after its inception, was a great success. The plant was running with about 70 workers, although the engineering estimate had called for 110. Costs were about 5 percent under those of conventional plants; daily absences were running below 1.5 percent, and turnover was less than 10 percent per year. Almost four years had passed without a single accident serious enough to cause absence. Workers' attitudes, which had been strongly positive almost from the beginning, remained so in most respects.

The experimental innovations that produced these results were numerous, although parts of a single coherent pattern. Supervisors are relatively few, and work groups of 7 to 14 members make day-to-day decisions about production, work allocation, and the like. Jobs rotate within these groups, also according to group decision, and workers are paid according to the number of jobs they have mastered rather than by the job they are doing during any particular hour or day. A worker who can do all the jobs in the plant earns the "plant rate," which is 50 percent above the base. These innovations in the division of labor, the distribution of power, and the allocation of rewards are complemented by the signs and symbols of organizational life, or rather by their absence. There is one entrance for all employees, one parking lot with no reserved places, one kind of carpeting throughout, from the workers' locker rooms to the plant manager's office.

Two major problems that became apparent in the Topeka Experiment as early as 1976 should be included in its assessment: the problem of renewal and the problem of diffusion. The workers and managers who began with Topeka had the unique experience of creating the new work environment; those who came after it had been well established were equal beneficiaries but not equal participants. This problem is in one sense absolutely insoluble; no generation can have the experience of its forbears.

To what extent the continued success of organizational experiments requires that the process of experimentation continue remains to be seen. It is certain, however, that workers' attitudes at Topeka, although they continued to be positive, became somewhat less so as the years passed.

The other problem at Topeka is its deviance; it was from the beginning an experimental plant in an otherwise conventional corporation. There are those who feel that such islands of deviance are unstable and that the corporation must become more like Topeka or Topeka must become more conventional. That question will surely be answered, but slowly. Meanwhile, the Topeka Experiment must be counted a complex and exciting success.

EXPERIMENTS IN SCANDINAVIA

Work experiments in Norway and Sweden are remarkable in several respects. They are numerous; in 1975, more than 500 were reported in Sweden alone (Lindholm, 1975). Many of them involve all three aspects of organizational structure—division of labor, distribution of power, and allocation of rewards. They are supported by trade unions as well as by management. The 1980 convention of Swedish labor unions put research on the quality of working life as its top priority (Gardell, personal communication), an event beyond all possibility in most countries. And in both Norway and Sweden national legislation has established standards for the quality of the work experience as well as for its physical demands and hazards. Those qualitative standards are not easily quantified or enforced, but they legitimize the redesign of work organizations and they make the goals of enhancing the work experience goals for the nation.

The best known of the Scandinavian experiments in the quality of working life involve the abolition of conventional assembly line methods in two automobile plants, Saab and Volvo. The Saab experiment initially included about 100 workers engaged in the final assembly of four-cylinder engines for passenger cars. After an extended period of planning by production and project development groups that included labor, management, technical experts, and union representatives, the "new" engine factory began production in 1972. The new factory was not a new building, but the space for assembling engines had been remodeled into bays, each of which could accommodate 10 workers. An ingenious carrier system brought the necessary parts to each bay, and an overhead conveyor brought in the engine blocks and removed the completed engines. The conveyor system did not set the pace of work, however; each bay accepted new blocks as they were needed, and sent the engines on their way when they were finished. Moreover, the division of labor within each bay was determined by the people in it. Each worker could assemble an entire engine, all 10 workers could set up a miniature assembly line, or they could establish some intermediate arrangement. When I observed the plant, most workers seemed to have chosen to work in pairs or small subgroups.

The Saab engines thus are produced not on an assembly line but by a number of parallel, relatively autonomous, self-paced work groups of 10 or fewer members. The system has been running since late in 1972, and neither management nor workers consider it an experiment any longer. Quantity of production is within the standards of conventional assembly methods. Absence and turnover, which had been special problems among assembly workers, are now no higher than in other worker categories. Work cycles, which would have been less than two minutes on a conventional assembly line of the same capacity, can be as much as 30 minutes, depending on decisions within the group. Survey data for 1976 (*Kilometeren*, December 1976) showed generally positive attitudes.

The Volvo plant at Kalmar, built in 1974, was truly new in concept, architecture, and technology. It was built to assemble automobiles in substantial numbers without a conventional assembly line. It is unlikely that a similar construction had occurred since Ford's introduction of large-scale assembly line methods 60 years earlier. Pehr Gyllenhammar (as quoted in Aguren, Hansson, and Karlsson, 1976) then the president of the Volvo company, was explicit about the objectives that the new plant had been designed to achieve:

> to organize automobile production in such a way that employees can find meaning and satisfaction in their work . . . [to] give employees the opportunity to work in groups, to communicate freely, to shift among work assignments, to vary their pace, to identify themselves with the product, to be conscious of responsibility for quality, and to influence their own work environment. (p. 5)

To a significant degree, most of those objectives have been met. The tasks of subassembly are performed in small groups, and the division of labor within each group is decided by its members, much as at Saab. The automobiles are carried from one such group to the next on electrically driven and guided platforms, which also permit workers to raise, lower, or turn over completely the automobile on which they are working. There are buffer zones between groups, so that the preferred pace of one group does not immediately set the pace for the next. Work cycles vary from 16 to 40 minutes. Employees are further involved in decisions through their several unions, all of which are represented in the plantwide works council, and through a number of "functional councils" for production, quality, personnel, and the like.

Worker perceptions and attitudes are positive but not indiscriminately so. About 9 of 10 workers report that teamwork is the practice within their group and that "job switching" (rotation) is practiced as well. Virtually all who do these things report that they like doing them. Most workers say that they exert some influence, both direct and indirect, over the way work is done. About half judge that influence to be small, however; only 25 percent rate their direct influence as "considerable," and 36 percent so rate their indirect influence.

These responses are more favorable than one would expect

from assembly line workers, either in the United States or in the older Volvo plants at Torslanda, which can serve as a basis of a comparison for the Kalmar plant. Compared to conventional plants, absence and turnover at the Kalmar plant are low, and by the standards of quantity, quality, and costs, the Kalmar plant equals conventional assembly methods. It does not, however, exceed them; Kalmar has increased the quality of working life, but cannot claim concomitant economic gains.

The Swedish experiments to improve work have taken place in a social context that has been strongly positive—supporting, encouraging, at times demanding changes in the organization of work, and almost always prompt to accept and legitimize those changes. That context includes a tradition of labor–management harmony and trade union strength, with 95 percent of blue-collar workers and 75 percent of white-collar workers belonging to unions. Unions have been represented on corporate works councils since 1971, and a series of agreements between union and employer confederations has expressed a joint commitment to experiments to improve the quality of working life. The government, as we have seen, has embodied in law the goals of such research.

The future of the Scandinavian experiments is now complicated by severe international competition, especially in manufactured goods. That competition is perhaps sharpest in automobile manufacture, with the Japanese industry setting the competitive pace. To what extent the improvement of working life can be made compatible with that pace remains to be discovered. One hopes that it may be and that the interests of producers and consumers may not seem increasingly divergent. In the long run they must converge, as findings of work-life experiments have also begun to converge.

Epilogue

We have now completed our examination of work and health. We have considered the nature of work in industrial society (Chapter 1) and the differences in job content that characterize different occupations (Chapters 2 and 3). We have identified the major stresses of work (Chapter 4) and the greater stresses of work deprivation (Chapters 5 and 6). We have raised the issue of goodness of fit between people and their work (Chapter 7) and reviewed the ways of changing organizations to improve goodness of fit (Chapters 8 and 9). Three large questions remain: What have we learned? What shall we do? How can we do it?

WHAT HAVE WE LEARNED?

We have learned that work has different meanings in people's lives. For some it is alienating and self-denying, a recurring investment of time and energy in activities that are unsatisfying or even feared and disliked. For others work is engrossing to the point of addiction, pursued to the detriment of health and the failed fulfillment of other life roles. For still others work offers self-fulfillment and satisfying relations with co-workers. Such contrasts involve differences among people, of course, but even more they involve differences among jobs.

In spite of such differences, work is important in the lives of most men and women. People answer questions about who they are by describing the work they do. Until they are very old, they say that they would prefer work to nonwork, and they value the friendships that arise in the work situation. It has been argued that people adjust to uninteresting work by the psychological trick of moving work from the center to the periphery of their lives, but the evidence tells us that work remains important. No other single activity demands so much time and energy. To set it aside is to call a great part of one's life wasted escept for the unavoidable getting of wages. But most people want more than money from their work, and in varying degrees they find it.

Social scientists usually study the nature of work by taking it apart, one characteristic at a time—how many hours, how much variety, how large a salary, and the like. But people do not have so many choices; work comes in packages, and when we get a job we get the whole package—demands and entitlements, good and bad. Occupations are the names that identify these packages, and there are many thousands of occupations. In spite of that complexity, people know very well which are the good jobs and what makes them good. Top government positions, the professions, and managerial or entrepreneurial jobs head the list. People who have such jobs are most satisfied and most likely to want to continue working. People who don't have such jobs may no longer aspire to them, but are likely to want them for their children or for themselves if they could start all over again.

The work people do and the way they feel about it are among the most important predictors of their overall well-being, and some of the other predictors (standard of living and character of nonworking activities, for example) are in large part work–determined. Satisfaction with work and satisfaction with life tend to go together.

The pragmatic question then becomes how to make work more satisfying. Not everybody can work in the preferred occupations, but perhaps we can design into all jobs some of those properties that make the preferred occupations so attractive. To do so we need to know what those properties are. Research directs us to eight basic aspects of any job: task content, autonomy and control, supervision and resources, relations with co-workers, wages, promotions, working conditions, and organizational context. Each of these can be further analyzed, as the following listing, adapted from the 1977 Quality of Employment Survey, suggests:

1. *Task content.*   Interesting work; the opportunity to use one's special abilities; able to see the results of one's work; a chance to do the things one does best.

2. *Autonomy and control.* Freedom to decide how to do one's own work; to choose the method; to set the pace.

3. *Supervision and resources.*   Enough information, help, and equipment to get the job done; clear responsibilities and enough authority to carry them out; a competent supervisor who is consistent and helpful, concerned about the welfare of those who report to him or her, and able to get people to work together.

4. *Relations with co-workers.*   Friendly and helpful co-workers; the opportunity to make friends with them.

5. *Wages and rewards.*   Good pay; job security; fringe benefits.

6. *Promotions.*   Promotions handled fairly; chances for promotion good; employer concerned about giving everybody a chance to get ahead.

7. *Working conditions.*   Enough time to get the job done; good hours; convenient travel; pleasant physical surroundings.

8. *Organizational context.*   Those properties of the organization as a whole that determine the quality of life within it: size, number of hierarchical levels, growth rate, competitive position, managerial philosophy, technology.

People's insights about what makes jobs good or bad correspond in many respects to the findings of stress research on what makes jobs healthy or unhealthy. The research on stress goes beyond the findings on job satisfaction, however, by linking occupations and job characteristics to objective physiological and medical outcomes. Stressful job characteristics include certain physical conditions (noise, crowding, bumping); punishment for failure, as contrasted with reward for successful performance; responsibility for the welfare of others; time pressure for task completion; short-cycle repetitive tasks; machine-paced work; sustained positional or postural constraint; and chronic overload.

All stress does not come from asking too much, however. Research both in the laboratory and in the work situation shows also the stressfulness of too little stimulation—too little variety in work and too little contact with others, for example. Still more stressful is the unhappy combination of both conditions, too much and too little, the performance of a small and repetitive repertoire to an exacting time schedule under constrained physical conditions. Physiological responses to such work include elevation of adrenaline and noradrenaline levels, elevation of blood pressure, gastric discomfort, and increased incidence of such diseases as hypertension, peptic ulcer, and diabetes.

The remedy for stressful work is not life without work. Research on unemployment, plant closings, and other forms of involuntary nonwork shows them to be more stressful than work. Unemployment, even with substantial economic benefits, is neither wanted nor well tolerated. Men and women in this society prefer to work. Despite the limitations of many jobs, no nonwork activity fulfills the same functions for most people, and most people know it.

We have learned, therefore, to think of work in terms of goodness of fit. Since unemployment is almost always stressful and alienating, and work is sometimes alienating but sometimes satisfying and fulfilling, the great question becomes how to improve the goodness of fit between individuals and their jobs.

That question leads us to a consideration of organizational change and the various strategies for improving the fit between personal needs and organizational requirements. The most widely used of those strategies I call strategies of adaptation, because they concentrate either on moving or changing individuals to fit organizational positions. Selection, placement, and training are the familiar elements in these strategies and all three of them have their places.

I am more optimistic, however, about strategies of coping, that is, ways of bringing human organizations more nearly into congruence with human needs and abilities. These coping strategies were further differentiated according to the aspect of organizational structure that they took as their primary target—the distribution of power, the allocation of rewards, or the division of labor. The most exciting experiments in improving the quality of working life by changing organizations seem to involve all three of these aspects, directly or indirectly.

WHAT SHALL WE DO?

According to the research reviewed, that question requires two answers, one having to do with the entitlement to work and the other with the quality of work life. First, we should make job entitlement a national commitment. Since private employers cannot always provide jobs for all those who need work and should be working, a national commitment to job entitlement implies that the government will be the employer of last resort. Many valuable products of the public work projects of the 1930s are still evident—from post offices and sidewalks to paintings and oral histories—and they remind us of the advantages of well-designed work programs.

A nation in which millions of people cannot find work is in danger, because the alternatives to work are often destructive, actively or passively. And the danger is increased when the line of cleavage that separates the employed from the unemployed coincides with other social boundaries, for example, those between old and young, black and white, or central cities and their surroundings. The availability of jobs, in short, is important for the health of both the individual and the nation.

Having a job, however, can be health giving or health damaging. The second answer to the question of what to do therefore involves the quality of jobs, no less than their availability. We have learned a great deal about what makes jobs good for people: challenge, the opportunity to work as part of a group, autonomy insofar as one's work is independent of others, and a share in decisions where others are involved. We have learned something, through much less, about how to create work organizations that offer such jobs. And we have learned that such organizational innovations and changes can often be had with maintained or enhanced productivity.

Improving the quality of working life is a goal that can be widely shared. It has the potential to bring together government and industry, unions and universities, employers and employees. It taps those human energies that, unlike other energy sources, are as renewable as life itself. It waits only to be given its place on the national agenda.

HOW SHALL WE DO IT?

The answers to that question, to the extent that we yet know them, have come from experience and research. To make those answers more complete, the research process must continue. Such research takes many forms, but it must include experiments in real-life work situations, on a substantial scale, and with extended evaluation. The stakes are too high and the costs of organizational change are too great for faddishness.

We can begin to improve the quality of working life on the basis of things already known, but to continue that process we must become an experimenting society (Campbell, 1969). That phrase does not imply scattered or casual change. To experiment is to test things worth trying, on a scale large enough to learn from and small enough so that we can afford to be wrong. It is a strategy that serves us well in science and in such applied scientific fields as medicine and engineering. It is the strategy of the clinical trial and the pilot plant. It can guide the search for better organizational forms and hasten the convergence of work and health.

References

Aguren, S., Hansson, R., and Karlsson, K. G. (1976) *The Impact of New Design on Work Organization*. Stockholm: The Rationalization Council SAF-LO.

Allardt, E. (1976). Work and political behavior. In R. Dubin, (Ed.), *Handbook of Work, Organization, and Society*. Chicago: Rand McNally.

Allport, F. H., (1933). *Institutional Behavior*. Chapel Hill: University of North Carolina Press.

Alphabetical Index of Industries and Occupations (1971). Washington, D.C., U.S. Bureau of the Census.

American Public Health Association (November 1975). *Health and Work in America: A Chart-Book*. Washington, D.C.: U.S. Government Printing Office (Available from the Superintendent of Documents).

American Society of Anesthesiologists (1974). Occupational disease among operating room personnel: A national study. *Anesthesiology*, **41**, 4.

Argyris, C. (1968). Some unintended consequences of rigorous research. *Psychological Bulletin*, **70** (3), 185–197.

Argyris, C. (1971). *Management and Organizational Development*. New York: McGraw-Hill.

Bakke, E. W. (1940). *Citizens without Work*. New Haven: Yale University Press.

Bardand, E. J. (1973). Results of current investigations of disability among meat wrappers in the Portland Metropolitan area. Eugene: University of Oregon Health Sciences Center.

Berman, L., and Goodall, M. C. (1971). Adrenaline, noradrenaline, and

185

3-methoxy-4-hydroxymandelic acid (MOMA) excretion following centrifugation and anticipation of centrifugation. *Federation Proceedings: Federation of American Societies for Experimental Biology,* **19**, 154. Cited by W. Raab in L. Levi (Ed.), *Society, Stress and Disease,* vol. 1, *The Psycho-Social Environment and Psychosomatic Diseases.* London: Oxford University Press.

Bowers, D. G. (1973). OD techniques and their results in 23 organizations: The Michigan ICL study. *Journal of Applied Behavioral Science,* **9**, 21–43.

Brayfield, A. H. and Crockett, W. H. (1955). Employee attitudes and employee performance. *Psychological Bulletin,* **52**, 396–424.

Brod, J. (1971). The influence of higher nervous processes induced by psycho-social environment on the development of essential hypertension. In L. Levi (Ed.), Society, Stress and Disease, vol. 1, *The Psycho-Social Environment and Psychosomatic Disease.* London: Oxford University Press.

Campbell, A., Converse, P. E., and Rodgers, W. L. (1976). *The Quality of American Life.* New York: Russell Sage Foundation.

Campbell, D. T. (1969). Reforms as experiments. *American Psychologist,* **24**, 409–429.

Caplan, R. D., Cobb, S., French, J. R. P., Jr., Harrison, R. D., and Pinneau, S. R., Jr. (1975). *Job Demands and Worker Health: Main Effects and Occupational Differences.* Washington, D.C.: U.S. Government Printing Office.

Cartwright, D. and Zander, A. (Eds.) (1960). *Group Dynamics: Research and Theory,* 2nd ed. Evanston, Ill.: Row, Peterson.

Cobb, S. (1969). In *Economics of Aging: Towards a Full Share in Abundance.* Hearings before the Subcommittee on Employment and Retirement Incomes of the Special Committee on Aging. U.S. Senate, 91st Congress, 1st Session, pp. 1199–1217. Washington, D.C.: U.S. Government Printing Office.

Cobb, S. (1974). Role responsibility: The differentiation of a concept. In A. McLean (Ed.), *Occupational Stress.* Springfield, Ill.: Thomas.

Cobb, S., and Kasl, S. (1977). *Termination: The Consequences of Job Loss.* U.S. Department of Health, Education and Welfare. Washington, D.C.: U.S. Government Printing Office.

Cohen, H. L. and Filipczak, J. (1971). A new learning environment. San Francisco: Jossey–Bass.

Cohen, W. J., Haber W., and Mueller, E. (1960). Impact of unemployment in the 1958 recession. Survey Research Center, University of

Michigan. Printed for the use of the Special Committee on Unemployment Programs. Washington, D.C.: U.S. Government Printing Office.

Conte, M., and Tannenbaum, A. S. (1977). *Employee Ownership.* A research study at the Institute for Social Research, University of Michigan, Ann Arbor.

Corson, S. A. (1971). Pavlovian and operant conditioning techniques in the study of psycho-social and biological relationships. In L. Levi, (Ed.), *Society, Stress and Disease,* vol. 1, *The Psycho-Social Environment and Psychosomatic Diseases.* London: Oxford University Press.

Counts, G. S. (1925). The social status of occupations, *School Review.*

Dean, R. (1966). Human stress in space. *Science,* **2,** 70.

Doll, R., and Jones, F. A. (1951). Occupational factors in aetiology of gastric and duodenal ulcers. *Medical Research Council,* Special Report Series, 276. London: HMSO.

Dubin, R. (Ed.) (1976). *Handbook of Work, Organization and Society.* Chicago: Rand McNally.

Dubin, R., Hedley, R. A., and Taveggia, T. C. (1976). Attachment to work. In R. Dubin (Ed.), *Handbook of Work, Organization, and Society.* Chicago: Rand McNally.

Dunn, J. P., and Cobb, S. (1962). Frequency of peptic ulcer among executives, craftsmen and foremen. *Journal of Occupational Medicine,* **4,** 343–348.

Fein, M. (1976). Motivation for work. In R. Dubin (Ed.), *Handbook of Work, Organization, and Society.* Chicago: Rand McNally.

Ferman, L. A. (1963). *Death of a Newspaper: The Story of the Detroit Times.* Kalamazoo, Mich.: W. E. Upjohn Institute for Employment Research.

Form, W. H. (1968). Occupations and careers. In D. L. Sills (Ed.), *International Encyclopedia of the Social Sciences,* vol. 11, New York: Macmillan and Free Press, pp. 245–254.

Frankenhaeuser, M. (1971). Experimental approaches to the study of human behavior as related to neuro-endocrine functions. In L. Levi (Ed.), *Society, Stress and Disease,* Vol. 1, *The Psycho-Social Environment and Psychosomatic Diseases.* London: Oxford University Press, pp. 22–35.

Frankenhaeuser, M., and Patkai, P. (1964). Inter-individual differences in catecholamine excretion during stress. *Scandinavian Journal of Psychology,* **6,** 117–123.

French, J. R. P., Jr. (1973). Person role fit. *Occupational Mental Health*, **3**, 15–20.

French, J. R. P., Jr. (1974). Person-role fit. In A. McLean, *Occupational Stress*. Springfield, Ill.: Thomas.

French, J. R. P., Jr., and Kahn, R. L. (1962). A programmatic approach to studying the industrial environment and mental health. *The Journal of Social Issues*, **18** (3) 1-47.

French, J. R. P., Jr., Rodgers, W. L., and Cobb, S. (1974). Adjustment as person-environment fit. In G. Coelho, D. Hamburg, and J. Adams (Eds.), *Coping and Adaptation*. New York: Basic Books.

Frost, C. F., Wakely, J. H., and Rhu, R. A. (1974). *The Scanlon Plan for Organizational Development: Identity, Participation and Equity*. Lansing: Michigan State University Press.

Gardell, B. (1976). *Job Content and Quality of Life*. Stockholm: Prisma.

Gosling, R. H. (1958). Peptic ulcer and mental disorder. *Journal of Psychosomatic Research*, **2**, 285.

Gross, N., Mason, W., and McEachern, W. A. (1958). *Explorations in Role Analysis: Studies of the School Superintendency Role*. New York: Wiley.

Haavio-Mannila, E. (1971). Satisfaction with family, work, leisure, and life among men and women. *Human Relations*, **24**

Hackman, J. R., and Suttle, J. L. (1977). *Improving Life at Work*. Santa Monica, Calif.: Goodyear.

Hamburg, D. A. (1962). Plasma and urinary corticoid levels in naturally occurring psychological stresses, in ultrastructure and metabolism of the nervous system. *Association for Research of Nervous Disease Processes*, **25**, 406.

Harrison, V. R. (1978). Person-environment fit and job stress. In C. L. Cooper and R. Payne (Eds.), *Stress at Work*. New York: Wiley.

Hatt, P. K., and North, C. C. Prestige ratings of occupations. In S. Nosow and W. H. Form (Eds.), *Man, Work, and Society*. New York: Basic Books.

Hebb, D. O. (1958). *A Textbook of Psychology*. Philadelphia: Saunders.

Herzberg, F., Mausner, B., Peterson, R. P., and Capwell, D. F. (1957). *Job Attitudes: Review of Research and Opinion*. Pittsburgh: Psychological Service of Pittsburgh.

Hodge, R., Siegel, P., and Rossi, P. (1965). Occupational prestige in the United States: 1925–1963. In R. Bendix and S. M. Lipset (Eds.), *Class, Status, and Power*. New York: Free Press.

Hollingshead, A. B., and Redlich, F. C. (1958). *Social Class and Mental Illness.* New York: Wiley.

House, J. S. (1972). The relationship of intrinsic and extrinsic work motivations to occupational stress and coronary heart disease risk. Doctoral dissertation, University of Michigan. *Dissertation Abstracts International,* **33**, 2514–A. (University Microfilms No. 72–29094).

Hunt, P., Schupp, D., and Cobb, S. (1966). An automated self-report technique. Unpublished manuscript. Ann Arbor, Mich.: Institute for Social Research.

Inkeles, A., and Rossi, P. H. (1956). National comparisons of occupational prestige. *American Journal of Sociology,* **61**, 329–339.

Institute of Medicine, National Academy of Sciences, 1981. Report of the Committee on Stress in Health and Disease.

Johnson, J. E. (1975). Stress reduction through sensation information. In I. G. Sarason and C. D. Spielberger (Eds.), *Stress and Anxiety,* vol. 2. Washington, D. C.: Hemisphere.

Katz, D., and Kahn, R. L. (1978). *The Social Psychology of Organizations,* 2nd ed. New York: Wiley.

Katz, D., Sarnoff, I., and McClintock, C. (1956). Ego-defense and attitude change. *Human Relations,* **9**, 27–45.

Komarovsky, M. (1940). *The Unemployed Man and His Family.* New York: Dryden Press.

Kornhauser, A. (1965). *Mental Health of the Industrial Worker.* New York: Wiley.

Kulka, R. A. (1975). Person and environment fit in U.S.: A validation study (2 vols.). Doctoral dissertation, University of Michigan. *Dissertation Abstract Institute,* 1976, **36**, 5252B. (University Microfilms No. 76–9438.)

Lawler, E. E. (1971). *Pay and Organizational Effectiveness: A Psychological View.* New York: McGraw-Hill.

Lazarsfeld-Jahoda, M. and Zeisel, H. (1933). *Die Arbeitslosen von Marienthal.* Leipzig: S. Hirzel.

Lazarus, R. S. (1966). *Psychological Stress and the Coping Process.* New York: McGraw-HIll.

Leavitt, H. J. (1965). Applied organizational change in industry: Structural, technological, and humanistic approaches. In J. G. March (Ed.), *Handbook of Organizations.* Chicago: Rand McNally.

Leinhart, W. S., Doyle, H. N., Enterline, P. E., Henschel, A., and Kendrick, M. A. (1969). *Pneumoconiosis in Appalachian Bituminous Coalminers.* Cincinnati: U.S. Department of Health, Education and Welfare.

Levi, L. (1972). Psychological and physiological reactions to and psycho-motor performance during prolonged and complex stressor exposure. In L. Levi (Ed.), *Stress and Distress in Response to Psycho-Social Stimuli.* Oxford: Pergamon Press.

Likert, R. (1961). *New Patterns of Management.* New York: McGraw-Hill.

Likert, R. (1967). *The Human Organization.* New York: McGraw-Hill.

Lindholm, R. (1975). *Job Reform in Sweden.* Stockholm: Swedish Employers' Confederation.

Lipset, S. M., Trow, J. A., and Coleman, J. S. (1956). *Union Democracy.* New York: Free Press.

Locke, E. A. (1976). The nature and causes of job satisfaction. In M. D. Dunnette (Ed.), *Handbook of Industrial and Organizational Psychology.* Chicago: Rand McNally.

Maslow, A. H. (1943). A theory of human motivation. *Psychological Review,* **50,** 370–396.

Maslow, A. H. (1954). *Motivation and Personality.* New York: Harper.

Maslow, A. H. (1970). *Motivation and Personality,* 2nd ed. New York: Harper & Row.

Meissner, M. (1971). The long arm of the job: A study of work and leisure. *Industrial Relations,* **10,** 238–260.

Menzies, I. E. P. (1960). A case study in the functioning of social systems as a defense against anxiety: A report of a study of the nursing services of a general hospital. *Human Relations,* **13,** 2.

Metzger, B. L. (1966). *Profit Sharing in Perspective.* Evanston, Ill.: Profit Sharing Research Foundation.

Michigan Organizational Assessment Package (1975). Progress Report 2. Ann Arbor: Institute for Social Research, University of Michigan.

Mirvis, P. H., and Berg, D. N. (1977). *Failures in Organizational Development and Change.* New York: Wiley-Interscience.

Morgan, J. (February 1981). Personal Communication.

Morse, N., and Reimer, E. (1956). The experimental change of a major organizational variable. *Journal of Abnormal and Social Psychology,* **52,** 120–129.

Morse, N., and Weiss, R. S. (1955). The function and meaning of work. *American Sociological Review,* **20** (2), 191–198.

Niez, J. (1935). The depression and the social status of occupations. *Elementary School Journal.*

Parker, S. R., and Smith, M. A. (1976). Work and leisure. In R. Dubin (Ed.), *Handbook of Work, Organization, and Society,* Chicago: Rand McNally.

Pflanz, M., Rosenstein, E., and Von Uexkull, T. (1956). Socio-psychological aspects of peptic ulcer. *Journal of Psychosomatic Research,* **1,** 68.

Porter, L. W., and Steers, R. M. (1973). Organizational work and personal factors in employee turnover and absenteeism. *Psychological Bulletin,* **80,** 151–176.

Quinn, R. P. (1977). *Effectiveness in Work Roles: Employee Responses to Work Environments,* vol. 1. Ann Arbor: Survey Research Center, University of Michigan.

Quinn, R. P., and Cobb, W., Jr. What workers want: Factor analyses of importance ratings of job facets. *The 1972-73 Quality of Employment Survey.* Ann Arbor: Institute for Social Research, University of Michigan.

Quinn, R. P., and Mangione, T. W. (1973). *The 1967-1970 Survey of Working Conditions.* Ann Arbor: Survey Research Center, University of Michigan.

Quinn, R. P. and Mangione, T. W. (1977). Fringe benefits: Some questions and answers. In R. P. Quinn et al. *The 1972-73 Quality of Employment Survey.* Ann Arbor: Survey Research Center, University of Michigan.

Quinn, R. P., and Shepard, L. J. (1972). *The 1972-73 Quality of Employment Survey.* Ann Arbor: Survey Research Center, University of Michigan.

Quinn, R. P., Walsh, T., and Hahn, D. L. K. (1977). The *1972-73 Quality of Employment Survey: Continuing Chronicles of an Unfinished Enterprise.* Ann Arbor: Survey Research Center, University of Michigan.

Raab, W. (1966). Emotional and sensory stress factors in myocardial pathology. *American Heart Journal,* **72,** 538.

Reiss, A. J., Jr. (1961). *Occupations and Social Status.* New York: Free Press.

Rice, A. K. (1958). *Productivity and Social Organization: The Ahmedabad Experiment.* London: Tavistock Publications.

Rice, A. K. (1963). *The Enterprise and Its Environment.* London: Tavistock.

Rioch, D. McK. (1971). The development of gastro-intestinal lesions in monkeys. In L. Levi (Ed.), *Society, Stress and Disease,* vol. 1, *The Psycho-Social Environment and Psychosomatic Diseases.* London: Oxford University Press.

Robinson, J., Athanasiou, R., and Head, K. (1969). *Measures of Occupational Attitudes and Occupational Characteristics.* Ann Arbor: Institute for Social Research, University of Michigan.

Roethlisberger, F. J., and Dickson, W. J. (1939). *Management and the Worker.* Cambridge, Mass.: Harvard University Press.

Russek, H. I., (1962). Emotional stress and coronary heart disease in American physicians, dentists and lawyers. *American Journal of Medical Science,* **243** (part 6), 716–725.

Russell, B. (1930). *The Conquest of Happiness.* New York: Liveright.

Selikoff, I. J., and Hammond, E. C. (1975). Multiple risk factors in etiology of environmental cancer: Implications for prevention and control. Unpublished data, cited in *Health and Work in America.* Washington, D. C.: U.S. Government Printing Office. Available from the Superintendent of Documents.

Seppanen, P. (1958). *Tehdas ja ammettiyhdistys.* (Dual allegiance to company and union.) Helsinki: WSOY.

Sheppard, H. L., Ferman, L. A., and Faber, S. (1960). Too old to work—too young to retire: A case study of a permanent plant shutdown. U.S. Senate. Special Committee on Unemployment Problems. Washington, D.C.: U.S. Government Printing Office.

Shrivastva, S. (1975). *Job Satisfaction and Productivity.* Cleveland: Case Western Reserve University Press.

Simon, H. A. (1974). How big is a chunk? *Science,* **183**, 482–488.

Skinner, B. F. (1967). *Science and Human Behavior.* New York: Free Press.

Skinner, B. F. (1974) *About Behaviorism.* New York: Knopf.

Slote, A. (1969). *Termination: The Closing at Baker Plant.* New York: Bobbs-Merrill.

Smith, P. C., Kendall, L. M., and Hulin, C. L. (1969). *The Measurement of Satisfaction in Work and Retirement.* Chicago: Rand McNally.

Sofer, C. (1972). *Organizations in Theory and Practice.* New York: Basic Books.

Srole, L., Langner, T. S., Michael, S. T., Opler, M. K. and Rennie, T. A. C., (1962). *Mental Health in the Metropolis: The Midtown Manhattan Study,* vol. 1. New York: McGraw-Hill.

Staines, G. L. (1977). Work and Nonwork: Part 1, A review of the literature. In R. P. Quinn (Ed.), *Effectiveness in Work Roles,* vols. 1 and 2. Ann Arbor: Survey Research Center, University of Michigan.

Staines, G. L. , and Pagnucco, D. (1977). Work and nonwork: Part 2, an empirical study. In R. P. Quinn (Ed.), *Effectiveness in Work Roles,* vols. 1 and 2. Ann Arbor: Survey Research Center, University of Michigan.

Stouffer, S. A. (1949). The American soldier. *Studies in Social Psychology,* Vols. 1 and 2. Princeton, N.J.: Princeton University Press.

Strauss, G. (1972). Is there a blue collar revolt against work? In J. O'Toole (Ed.), *Work and the Quality of Life.* Cambridge: Massachusetts Institute of Technology Press.

Tannenbaum, A. S. (1974). *Hierarchy in Organizations.* San Francisco: Jossey-Bass.

Tannenbaum, A. S., and Allport, F. H., (1957). Personality structure and group structure: An interpretive study of their relationship through an event structure hypothesis. *Journal of Abnormal and Social Psychology,* **53**, 272-280.

Tannenbaum, A. S. and Cooke, R. A. (1978). Organizational control: A review of research employing the control graph method. In C. J. Lammers and D. C. Hickson (Eds.), *Organizations Alike and Unlike.* London: Routledge & Kegan Paul.

Tannenbaum, A. S., and Kahn, R. L. (1958). Leadership practices and member participation in local unions. *AFL-CIO Educational News & Views,* **3** (4), 5–6.

Taylor, F. W., (1911). *The Principles of Scientific Management.* New York: Harper.

Taylor, J. C. and Bowers, D. C. (1972). *Surveys of Organizations.* Ann Arbor: Institute for Social Research, University of Michigan.

Terkel, S. *Working.* New York: Random House.

Trist, E. (1976). Toward a postindustrial culture. In R. Dubin (Ed.), *Handbook of Work, Organization and Society.* Chicago: Rand McNally.

Trist, E. L., and Bamforth, K. W., (1951). Some social and psychological consequences of the long-wall method of coal-getting. *Human Relations,* **4**, 3–38.

Trist, E., Higgin, G. W., Murray, H., and Pollock, S. B. (1963). *Organizational Choice.* London: Tavistock Publications.

Vertin, T. G. (1974). Cited by S. Cobb. Role responsibility: The differentiation of a concept. In A. McLean (Ed.), *Occupational Stress.* Springfield, Ill. Thomas.

Vroom, V. H. (1960). *Some Personality Determinants of the Effects of Participation.* Englewood Cliffs, N.J.: Prentice-Hall.

Weiss, R. S. , and Kahn, R. L. (1960). Definitions of work and occupation. *Social Problems,* **8,** (2), 142–151.

Whyte, W. F. (1972). Skinnerian theory in organizations. *Psychology Today,* **5** (11), 66.

Wilensky, H. (1964). Varieties of work experiences. In H. Borow (Ed.), *Man in a World of Work.* Boston: Houghton Mifflin.

Zung, W. K. (1955). A self-rating depression scale. *Archives of General Psychiatry,* **13,** 63–70.

Index

DATE DUE